THE BIG BANG

A HISTORY OF EXPLOSIVES

A gigantic explosion from a drawing by Leonardo da Vinci. (The Royal Collection © Her Majesty The Queen)

THE BIG BANG

A HISTORY OF EXPLOSIVES

G.I. BROWN

FOREWORD BY
ADAM HART-DAVIS

SUTTON PUBLISHING

First published in 1998 by
Sutton Publishing Limited · Phoenix Mill
Thrupp · Stroud · Gloucestershire · GL5 2BU

Paperback edition first published 1999

Reprinted 2000, 2001

British Library Cataloguing in Publication Data
A catalogue record for this book is available from the British Library

ISBN 0 7509 2361 X

Typeset in 11/12pt Ehrhardt.
Typesetting and origination by
Sutton Publishing Limited.
Printed and bound in England by
J.H. Haynes & Co. Ltd, Sparkford.

Contents

For Barbara, Helen and Louise

Foreword

When George Brown sent me the typescript of this book, my heart sank a little. I had plenty of work to do at the time, and the thought of having to read a treatise on the history of explosives filled me with gloom. However, feeling I should be polite, I started reluctantly at page 1 – and by page 3 found I was utterly hooked. I won't say I couldn't put it down, because in fact it was in a huge ring-binder that was so heavy I could scarcely pick it up, but I found the book completely fascinating. I read it right through, and now wait impatiently for a bound copy to read again, with all the excellent illustrations in place.

The first few pages give a fine flavour of *The Big Bang*, with the intriguing story of Greek fire, and how it was ousted by the new-fangled gunpowder. The struggles of so many people to make good gunpowder, the secret code of Roger Bacon, and the enigma of Black Berthold Schwartz had me on the edge of my seat.

I present a television series called *Local Heroes*, which celebrates pioneers of science, technology and invention, so I am always interested in stories about how people have ideas and introduce new technology. *The Big Bang* is full of such stories: in particular I recommend those of Bickford, Forsyth and Congreve. William Bickford, a Cornish currier in Tuckingmill, was so distressed by seeing the results of accidents with blasting powder in mines that he invented the safety fuse, which has scarcely changed till this day. Scottish minister Alexander Forsyth was frustrated when he tried to shoot ducks on the loch near his house at Belhevie north of Aberdeen, because the ducks saw the flash in the pan of his flintlock gun and flew away before the shot arrived; so he invented the percussion cap, which transformed military guns. Yet neither of these two made any money before they died. Colonel Sir William Congreve, by contrast, was much more successful with his rockets – even though they were more effective at terrifying horses than at damaging the enemy. However, he did manage to set fire to Boulogne by mistake, and his rockets blazed their way into the American National Anthem.

George Brown has done an exceptional job of weaving together the strands of history, biography, politics and science. I was fascinated to learn about the du Ponts, Alfred Nobel and Robert Oppenheimer; about Guy Fawkes and the Gunpowder Plot, about Count von Rumford and his extraordinary love-life, and about Chaim Weizmann and the national conker collection. I am most grateful to George for bringing these stories to my attention, and I hope you too will enjoy this splendid book.

Adam Hart-Davis
May 1998

Acknowledgements

The author would like to thank the following individuals and organizations for their help:

Argonne National Laboratory, USA; BASF; British Museum; Calladines (Stationery), Eton; Chemistry in Britain (The Royal Society of Chemistry); the librarian of Christ Church College, Oxford; Compair Holman Ltd; DuPont (UK) Ltd; Eley Hawk Ltd; the librarians of Eton College; Explosive Storage and Transport Committee of the Ministry of Defence; Faversham Society, Kent; Federation of Demolition Contractors; Greater Manchester County Record Office; Hagley Museum and Library, USA; Health and Safety Laboratory (Explosives Section), Buxton; Peter Huggins, ICI; ICI Australia Operations Pty Ltd; Imperial War Museum; Italesplosivi, Milan, Italy; Julian Cleeton; G.N.G. Tingey; Los Alamos National Laboratory, USA; National Maritime Museum; National Railway Museum; Oxford University Press; The Patent Office; Quarry Products Association; Robinson & Birdsell Ltd, Leeds; Royal Air Force Museum, Hendon; Royal Armouries, Leeds; Royal Artillery Historical Trust; Royal Collection Enterprises; Royal Commission on the Ancient and Historical Monuments of Scotland (RCAHMS); and the librarians of the Science Museum.

He is also grateful for help from Anne Marshall, Adam Hart-Davis and John Briscoe; to a number of friends who read the manuscript in its early days; and to his patient wife. She was a constant source of encouragement and provided both historical background and useful reminders when the technical matters became too much for her.

CONVERSION FACTORS

1 millimetre (mm)	= 0.039 inch			
1 metre (m)	= 1000 mm	= 1.094 yard	= 3.2808 feet	= 39.37 inch
1 kilometre (km)	= 1000 m	= 0.621 mile	= 1093.6 yard	= 3280.8 feet
1 gram (g)	= 0.035 ounce			
1 kilogram (kg)	= 1000 g	= 2.205 pound	= 35.274 ounce	
1 Megagram (Mg)	= 1000 kg	= 1 tonne	= 0.984 ton	= 2204.6 pound

1 atmosphere (atm) = 101,325 Pascal (Pa) = 14.7 pound per square inch

CHAPTER 1
The Invention of Gunpowder

Fire can cause so much damage and create so much confusion, chaos and terror that it is not surprising that the use of incendiaries in warfare dates back to very ancient times. Fire-pots being thrown down on troops besieging a town are depicted in Assyrian bas-reliefs in the British Museum, dated around 900 BC, and early writers give detailed, and sometimes lurid, accounts of the use of incendiaries. Herodotus describes how arrows tipped with burning tow were used in the capture of Athens in 480 BC, and Thucydides tells how a huge bonfire was lit against the wooden walls of Platea[1], in 429 BC. Even more remarkably, he describes the use of a blow-pipe in the attack on Delium in 424 BC:[2]

> They took a great beam, sawed it in two parts, both of which they completely hollowed out, and then fitted the two parts closely together again, as in the joints of a pipe. A cauldron was then attached with chains to one end of the beam, and an iron tube, curving down into the cauldron, was inserted through the hollow part of the beam. Much of the surface of the beam itself was plated with iron . . . When this machine was brought up close to the city wall, they inserted into their end of the beam large bellows and blew through them. The blast, confined inside the tube, went straight into the cauldron which was filled with lighted coals, sulphur and pitch. A great flame was produced which set fire to the wall and made it impossible for the defenders to stay at their posts. They abandoned their positions and fled; and so the fortification was captured.

At first, almost any locally available materials which would burn were used as fuel but, with time, an unusual degree of sophistication crept in. A collection of supposedly contemporary recipes[3] reveals a whole catalogue of witches' brews. 'Another kind of fire for burning enemies wherever they are can be made', we are told[4], 'by taking petroleum, liquid pitch, and oil of sulphur. Put all these in a pottery jar buried in horse manure for fifteen days. Take it out and smear with it crows which can be flown against the tents of the enemy. When the sun rises and before the heat has melted it the mixture will inflame. But we advise that it should be used before sunrise or after sunset.' In Arabian manuscripts, the unfortunate crows were apparently set on fire before launching; similar use of fire-birds appears in early Chinese works.

There were many other horrible formulations, all trying to outburn the others, but the use of incendiary materials entered a new phase in AD 673 when an architect called Kallinikos carried the secret of Greek fire from Heliopolis to Constantinople. That majestic city was the centre of the Byzantine Empire and Greek fire was to be the lynch-pin of the empire's successful defence, against attacks from all quarters, for the next 800 years. It was an embryonic napalm, the Palladium of the Empire and, without doubt, the greatest deterrent of the day.

But quite what Greek fire was must remain something of a mystery. The name itself is widely used, or misused, for almost any incendiary mix, and may be used synonymously with wild-fire, wet-fire, sea-fire, maritime-fire, automatic-fire or incendiary oil. In Britain, where the incendiary was first introduced towards the end of the twelfth century, the term wildfire was generally preferred, and has remained in usage in the phrase 'to spread like wildfire'. The fact that Greek fire did spread very rapidly over the surface of water gives some indication of its nature. It was always described as a liquid or semi-liquid material; it could be propelled through tubes; it floated on water; it was very difficult to extinguish; and it was possible, somehow, to keep the secret of its manufacture within Constantinople for many years. The secret was, indeed, very well protected by the Official Secrets Act of the day. The Emperor Constantine VII decreed that enquirers were to be told that it had been revealed by an angel, and there were very stern warnings that any communication of it to the enemy was treasonable and sacrilegious, bringing with it divine retribution. Someone about to betray the secret, it was said, had been struck by lightning.

What was the secret? All the evidence suggests that the main component of Greek fire was made from rock-oil or naphtha. This is a black, sticky material which does not occur at all commonly on the earth's surface but which is found, oozing out of the ground or floating on pools of water, in the region between the Black Sea and the Caspian Sea. Rock-oil does not burn very easily, nor is it a very good solvent. But, if it is distilled – and the great secret was probably how to do this safely – a lighter, more inflammable oil can be obtained. It was probably an oil like this, thickened by dissolving in it such substances as sulphur and resin, that made up Greek fire. An early recipe[5] says 'You will make Greek fire in this way. Take liver of sulphur, tartar, sarcocolla and pitch, boiled salt, petroleum oil and common oil. Boil all these well together. Then immerse in it tow and set it on fire. If you like you can pour it through a funnel. Then kindle the fire, which is not extinguished except by urine, vinegar or sand.'

On land, Greek fire was either hurled from ballistae in large tubs with the mixture already alight, or projected in pots and then ignited by incendiary arrows. At sea, it was ejected through tubes or siphons. An account of a naval battle between the Greeks and the Pisans in 1103 gives a good, general picture but is very short on technical detail. 'Each of the Byzantine galleys', it says, 'was fitted in the prow with a tube ending with the head of a lion or other beast made of brass or iron, and gilded, frightful to behold, through the open mouth of which it was arranged that fire should be projected by the soldiers through a flexible apparatus.'[6] It is not at all clear how all this could have been achieved in practice. Playing with fire is a dangerous business and the player can easily get his own fingers burnt. Nevertheless, the problems must have been overcome successfully, because Greek fire served the Byzantines extremely well. The empire flourished,

Greek fire being used at sea. A drawing from a tenth-century Byzantine manuscript. (Reprinted from Fire *by Hazel Rossotti (1993), by permission of Oxford University Press)*

in all, for eleven centuries and only fell in 1453, when the Greek fire was overpowered by the relatively newfangled gunpowder of the Turkish invaders.

* * *

For many hundreds of years, until well into the second half of the nineteenth century, gunpowder was the only explosive to be used both as a propellant in guns and for blasting in warfare, mining and civil engineering. It only began to lose its monopoly when so-called nitro-explosives, such as dynamite and guncotton, were discovered around 1850.

As it looks rather like soot, gunpowder is commonly called black powder. A typical modern sample contains 75 per cent of potassium nitrate (nitre or saltpetre), 10 per cent of sulphur and 15 per cent of wood charcoal (carbon), but early mixtures contained much smaller amounts of saltpetre. All three components must be finely powdered, and they must be well mixed.

Sulphur is a yellow solid which will burn in air with a rather feeble blue flame. Wood charcoal simply smoulders, as in a barbecue. Both will burn much more rapidly in an oxygen-rich atmosphere, particularly if they are finely powdered, and it is the potassium nitrate that holds the key to gunpowder because it provides the oxygen. When gunpowder, or black powder, is lit, the oxygen from the nitrate enables the sulphur and the carbon to burn rapidly, forming a mixture of hot gases containing mainly sulphur dioxide and carbon dioxide, so that there is a large, and rapid, increase in volume. But this will only lead to an explosion if the gunpowder is lit in a confined space such as a sealed container, a gun barrel or a borehole, so that the pressure can build up sufficiently.

The pressure rise with gunpowder is never as dramatic as it is with high explosives, such as dynamite or guncotton, and it is classified as a low explosive. It is not very effective in cutting through steel plate, but it is suitable for pushing projectiles out of gun barrels, or blasting rock when it is set off in a borehole. It is referred to as a propellant explosive to distinguish it from high explosives which have a much greater shattering effect. Dynamite delivers a knock-out blow; gunpowder gives a more prolonged, steady push.

Gunpowder is easy to handle, easy to set off (it only requires to be heated above 300°C) and very safe, so far as any explosive can be regarded as safe. If lit in the open it simply burns away very quickly, even in the absence of air. The major disadvantages are its lack of power, its production of a lot of nasty smoke and fumes when it explodes, and its failure to explode when it is damp. In the early days, setting off gunpowder always involved some sort of heat or fire – a hot rod, a flame, a spark or frictional heat – and the use of the word 'fire' has persisted in 'fireworks', 'gunfire' and 'firing the gun', even though modern guns are generally set off by the explosion of a very sensitive chemical by a sharp blow, or percussion, as in a child's toy cap-gun.

* * *

Gunpowder is both the best known and the oldest explosive, but its origins are shrouded in some mystery. Any researcher is soon entangled in a web of mistakes, misinterpretations and misrepresentations and the chance of finding a definitive answer has always been rather bleak. But a few historians have continued the search, sifting through claims that gunpowder was first used in Persia, in India, in Arabia, and in China. The most authoritative modern view is that gunpowder was first made in China, in the middle of the ninth century AD, by Thang alchemists who were actually looking for the elixir of immortality – one of the most remarkable examples of inventing one thing while trying to find something else. The early Chinese literature refers to 'fire-chemical' and 'fire-drug',[7] but it is not until 1004 that there is any specific mention of the composition of gunpowder and even then no information is given about the proportions in the mixture.

The earliest use of gunpowder was certainly in fireworks, for which the Chinese have always had a passion, but the possibility of making simple bombs and grenades was realized during the eleventh century. The military implications led the Chinese to place the production of sulphur and saltpetre under state control and in 1067 the emperor banned their sale to foreigners. The Chinese army was already well over a million men strong and it was necessary to arm it as well as possible to fend off attacks by the Mongols from central Asia.

No one knows with any certainty how the news of all this remarkable activity in the East reached the Western world, nor why it travelled so slowly, but it may have been carried by the Saracens – those middle-men between the Orient and the West – for Roger Bacon, who first told the tale in about 1260, could read Arabic.

* * *

A drawing of Roger Bacon by Michael Ayrton. (ICI)

Roger Bacon was born in 1214 at Ilchester in Somerset. An inscription placed in the Church of St Mary Major there 'by a few admirers of his genius' to commemorate the seventh centenary of his birth gives a sketch of his life and achievements. It reads:

> To the immortal Memory of Roger Bacon, a Franciscan Monk and also a free enquirer after true knowledge. His wonderful powers as mathematician, mechanician, optician, astronomer, chemist, linguist, moralist, physicist, and physician gained him the title of Doctor Mirabilis. He first made known the composition of gunpowder, and his researches laid the foundations of modern science. He prophesied the making of machines to propel vessels through the water without sails or oars; of chariots to travel on land without horses or other draught animals; of flying machines to traverse the air. He was imprisoned, starved and persecuted by the suspicious ignorance of his contemporaries, but a fuller knowledge now acclaims and honours him as one of the greatest of mankind. Born at Ilchester in 1214. Died at Oxford in 1292.

The sentiments, as in many church inscriptions, may be a little fulsome, but there can be no doubt that Roger Bacon was a very learned and very unusual man, though he remains something of an enigma to historians. Some rate him as heralding the dawn of modern science almost single-handed amid the gloom of the thirteenth century; others have regarded him as a sorcerer and necromancer whose reputation hangs only on legend.

Little is known of his boyhood but he came from a wealthy family who, later in his life, sacrificed their fortunes to help King Henry III in his struggle with the barons. Roger was a pupil of Robert Grosseteste at Oxford, where he studied theology, geometry, arithmetic, music, astronomy, Greek, Latin and Arabic. After some time at the University of Paris, where he graduated so brilliantly that he was nicknamed 'Doctor Mirabilis', he returned to Oxford and became a Franciscan friar around 1250. He lived at a time when learning, independence and freedom of speech were not greatly encouraged, but when a new outlook was being forced on to Western Christendom by the impact of the recently imported translations of major works of Greek and Arabian philosophy.

Bacon, in common with all other Christians of his day, believed that the Bible contained, in one form or another, the whole realm of knowledge. In the preface to his *Opus Maius*, written at Pope Clement IV's request in an attempt to reassess the changing situation, he states: 'I wish to show that there is one wisdom which is perfect, and that this is contained in the scriptures. From the roots of this wisdom all truth has sprung. I say therefore that one science is mistress of all others, namely theology.'[8] Yet within that medieval concept, he was a great advocate of careful observation and experiment; what would, nowadays, but in a very different context, be called experimental method. He regarded mathematics as the 'gateway and key to all other sciences' and he hoped to demonstrate that natural science, far from being a danger to Christendom, was a source of wisdom and power.

It was the beginning of the Science versus Religion controversy. In all his searchings for the 'perfect wisdom', Bacon applied his famous adage 'sine experimentia nihil sufficienter sciri potest' ('nothing can be certainly known but by experience'). But such a new and independent outlook – not always expressed with great tact – brought him into conflict with other Franciscans, some of whom he attacked as conceited and corrupt pedants. So much so that, in the end, his writings and his freedom were restricted. One report says that Jerome of Ascali, the Minister General of the Franciscans in 1277, 'by the advice of many friars, condemned and denounced the teaching of Roger Bacon of England, master of sacred theology, as containing some suspected novelties, on account of which the said Roger was condemned to prison, with the order given to all the brethren that none should hold his doctrine but avoid it as reprobated by the Order'.[9]

* * *

Bacon revealed the composition of gunpowder in a treatise entitled *De secretis operibus artis et naturae et de nullitate magiae* ('On the marvellous power of art and of nature and on the nullity of magic'). It consists of eleven letters or chapters, and versions of it are known in Latin, French, German and English, though their validity and the dates of their publication have been much argued by scholars. The work attempts to prove that certain happenings, which at the time were attributed to evil magic, could be due to natural causes and could be imitated by experiment, and it was probably written to defend the author against the charge that he was guilty of magic. Bacon suggests that such unheard-of possibilities as submarines, aeroplanes, compasses, motor cars, suspension bridges and magic lanterns could all be made to exist by natural means and without any trickery at all. This was far advanced for 1260. Yet he is at pains not to reveal any secrets easily. He writes: 'But I recall that secrets of nature are not to be committed to the skins of goats and of sheep [that is, vellum and parchment] that anyone may understand them', and 'a man is crazy who writes a secret unless he conceals it from the crowd and leaves it so that it can be understood only by the effort of the studious and the wise'. And, after explaining seven different ways of encoding secrets, he adds 'I have judged it necessary to touch upon these ways of concealment in order that I may help you as much as I can. Perhaps I shall make use of them because of the magnitude of our secrets.'[10]

One of the secrets to be revealed was the composition of gunpowder. Bacon refers to a material by which 'the sound of thunder may be artificially produced in the air with greater resulting horror than if it had been produced by natural causes'. He then gives cryptic accounts of how to purify 'aerial stone' or 'stone of Tagus' (meaning saltpetre) – a vital part of making gunpowder; refers to 'certain parts of burnt shrub or willow' (charcoal); and to 'vapour of Pearl' (sulphur). But the most famous passage in the book, not universally accepted as authentic, reads as follows:

Sed tamen salispetrae LURU VOPO VIR CAN UTRIET sulphuris et sic facies tonitruum et coruscationem; sic facies artificium; Vides tamen utrum loquor in seneigmate, vel secundun veritamet. ['But however of saltpetre LURU VOPO VIR CAN UTRIET of sulphur, and so you will make thunder and lightning, and so you will make the artifice (or turn the trick). But you must take note whether I am speaking in an enigma or according to the truth.'][11]

The phrase in capitals is unintelligible and it seems likely that the words at the end of the chapter[12] – 'Whoever will rewrite this will have a key which opens and no man shuts; and when he will shut, no man opens' – applied to it. The key remained hidden for almost 650 years but it was deciphered, in 1904, by Lt-Col. Hime, a Royal Artillery officer who made a study of the history of Greek fire and gunpowder. He recognized the letters as an anagram and rearranged them as follows

R VII PART V NOV CORUL V ET

A version of Roger Bacon's gunpowder anagram. (British Museum, SLOANE MS. 2156)

so that the recipe for gunpowder follows as: 'take [R still means take in today's doctors' prescriptions] seven [of saltpetre], five of young hazelwood and five [of sulphur]'. This gives a gunpowder containing 29.4 per cent each of charcoal and sulphur and 41.2 per cent of saltpetre, which is very close to the composition of early gunpowder mixtures.

One can only guess why Bacon hid the details so carefully. Perhaps he was frightened of publicizing such a revolutionary and awesome product which might well be regarded as associated with magic; perhaps he feared the reaction of the Church and was all too aware of the recently founded Inquisition; perhaps he did it all in jest. The puzzle is all the odder given that within a few years he was writing much more openly about gunpowder. In the *Opum Tertium*, about 1267, he wrote: 'There is a child's toy of sound and fire made in various parts of the world with powder of saltpetre, sulphur and charcoal of hazelwood. This powder is enclosed in an instrument of parchment the size of a finger, and since this can make such a noise that it seriously distresses the ears of men, especially if one is taken unawares, and the terrible flash is also very alarming, if an instrument of large size were used, no one could stand the terror of the noise and flash.'[13]

There can be no doubt that Bacon knew quite a lot about gunpowder, however secretly he may have imparted his information, but the claim, sometimes made, that he invented it is certainly false. It had been known in China for almost 400 years before he wrote about it in the West.

* * *

Some even doubt Bacon's claims to be the first to bring the news of gunpowder to the West, for other scribes were active around the same time. Count Albert of Bollstadt, better known as Albertus Magnus, and sometimes as Saint Albert the Great, was born in Lauingen, on the Danube, around 1200 and he lived until 1280. He, like Bacon, was a great scholar and wrote thirty-eight volumes on almost every aspect of contemporary knowledge, and he gave the composition of gunpowder in *De Mirabilis Mundi* ('Marvels of the World'). Almost the same recipe also appears in *Liber Ignum* ('Book of Fires'), a six-page pamphlet written about 1225 and generally attributed to Mark the Greek (Marcus Graecus), but little is known about who he might have been, even if he existed at all, and it is probable that he copied from Albertus Magnus, or vice versa.

Berthold Schwartz, commonly known as Black Berthold, is another mysterious figure on the scene. He is commemorated by a statue in Freiberg, which was an important centre for casting cannon and training gunners in the fourteenth and fifteenth centuries, and is portrayed in contemporary prints. The statue claims that he invented gunpowder and guns in 1353 and one copper engraving depicts him firing a charge of gunpowder with a flint of steel and has an inscription describing him as 'worthy and ingenious' and as a 'Franciscan Monk, doctor, alchemist and inventor of the force of using firearms in the year 1380'. A less complimentary epitaph had him as the 'most abominable of inhumans who, by his art, had made miserable all the rest of humanity'.[14] Yet though he was obviously buried and revered with some ceremony and feeling, it is not clear whether he

Gunpowder exploding in a mortar while being tested by Berthold Schwartz. (An engraving by R. Custos, 1643)

ever actually lived. Professor Partington, who investigated Schwartz's claims in great depth, added him to the list of legendary figures alongside William Tell and Friar Tuck.

* * *

So, despite much very careful work by a few dedicated researchers, the questions 'What is Greek fire?' and 'Who invented gunpowder?' still cannot be answered with any certainty. Let Edward Gibbon sum it up in his elegant, but vague, eighteenth-century judgements. On Greek fire, he wrote: 'The historian who presumes to analyse this extraordinary composition should suspect his own ignorance and that of his Byzantine guides, so prone to the marvellous, so careless, and in this instance, so jealous of the truth!'[15] and on gunpowder, 'the precise era of the invention and application of gunpowder is involved in doubtful tradition and equivocal language'.[16]

CHAPTER 2

Making Gunpowder

Describing gunpowder in the seventeenth century, John Bate[1] wrote that 'the saltpetre is the Soule, the Sulphur is the Life and the Coales the Body of it'. Saltpetre, or nitre, is certainly the most important component. It was recognized as potassium nitrate by John Mayow, a friend of Robert Boyle's, in 1674. Prior to that much confusion had been caused by using the word 'nitre' to mean what we now know as sodium carbonate. The difficulty arose because the various different salts known to the ancients were all remarkably alike in both appearance and taste. The biblical quotations in Jeremiah ii 22 – 'For though thou wash thee with nitre, and take thee much soap, yet thine iniquity is marked before me, saith the Lord God' – and in Proverbs xxv 20 – 'As he that taketh away a garment in cold weather, and as vinegar upon nitre, so is he that singeth songs to an heavy heart' – only make sense if nitre is interpreted as sodium carbonate. And Robert Boyle demonstrated in 1680 that a sample of Egyptian nitre, given to him by the British ambassador at the Ottoman court, was a carbonate and not a nitrate. He wrote: 'When once I received the nitre that I have mentioned, I quickly poured upon it some vinegar, and found as I expected that there presently ensued a manifest conflict, with noise, and store of bubbles, with which experiment I afterwards acquainted some critics and other learned men who were not ill-pleased with it.'[2] Around that time the Egyptian carbonate came to be called natron, and the word nitre was applied to potassium nitrate. In modern versions of the Bible, the term nitre is replaced by lye or soda.

The word saltpetre is derived from the Latin *sal petrae*, salt of stone, the name originating from the fact that the chemical was first encountered as a white encrustation on stones or walls or on the ground. In Egypt it was called 'the flower of the stone of Assos', or 'Chinese snow',[3] as it looks like snow on the ground, but these names are no longer in use.

Until the middle of the nineteenth century all the saltpetre required for making gunpowder was obtained by dissolving it from earth in which it had formed naturally by the decomposition of animal and vegetable matter. It is now known that this decomposition involves the action of bacteria, and plentiful supplies of saltpetre can only build up where there is an abundance of nitrogenous organic matter in the earth, where the temperature is high so that the bacterial action is accelerated, and where there is a prolonged dry season so that the saltpetre does not get washed away as soon as it is formed. Alternatively, saltpetre can build up in cellars and stables sheltered from rain. Wherever it occurred in large quantities

Sixteenth-century nitre beds from a treatise by Lazarus Ercker, 1580. Wood ashes, from the wood piles (D), were added to nitrate-containing earth scraped from the beds (C). The mixture was treated with water in the leaching house (A) to extract the saltpetre. The solution was then heated in the boiling house (B) until it formed crystals on cooling.

it was sought after like gold, for it was imperative for every country to have adequate supplies, particularly in time of war.

India, especially the states of Bihar and Bengal, was one of the largest suppliers, exporting around 30,000 tons of saltpetre each year in the first half of the nineteenth century. The rapid nitrification of the sewage-sodden earth around the villages and the material from the mud walls of houses or cow sheds, produced a crust very rich in nitrates. The nitrate-containing earth was collected into a pile, and then wood ashes, which contain potassium carbonate, were mixed in so that any calcium or magnesium nitrates which might be present would be converted into the required potassium nitrate, and water was allowed to trickle through. The solution which was collected was, subsequently, evaporated either by a fire or by the sun's heat, to form crystals.

The first crop of crystals was not very pure but could be refined by recrystallization to yield a solid containing about 90 per cent potassium nitrate. Further purification, particularly the removal of deliquescent impurities, was essential before the saltpetre could be used in gunpowder, for deliquescent substances absorb moisture from damp air. This necessity had been recognized in an early test given in the *Codex Germanicus*, written around 1350: 'When thou buyest or makest saltpetre and will find whether it be good or not, so thrust thine hand there-into. If thine hand become damp, then it is not good. Also touch thine hand with thine tongue; if thine hand be salty, then the saltpetre is not good; but if thine hand be sweet, then it is good.'[4] Early methods of purification had also been described by Roger Bacon and by Marcus Graecus, but considerable chemical ingenuity, making use of the differences in salt solubilities, had to be applied before an entirely satisfactory method was developed.

* * *

The basic conditions for the formation of nitrates in the ground were considerably less favourable in Europe than in India so that, in spite of strenuous and sometimes rather ridiculous governmental efforts, home production of saltpetre in Europe was always difficult.

In England early saltpetre manufacture was controlled by the State. In 1558 Queen Elizabeth I granted an eleven-year monopoly for gathering and making the material to George Evelyn (grandfather of John Evelyn, the famous diarist) and others. The monopoly covered all of southern England and the Midlands, excepting the City of London and its environs. By 1625 King Charles I had empowered saltpetre makers to enter any premises to remove nitrate-containing earth wherever they could find it and, in the following year, he went even further, ordering that his

> loving subjects . . . inhabiting within every city, town and village . . . shall carefully and constantly keep and preserve all the urine of man during the whole year, and all the stale of beasts which they can save and gather together whilst their beasts are in their stables and stalls, and that they be careful to use the best means of gathering together and preserving the urine and stale, without mixture of water or other thing put therein. Which our commandment and royal pleasure being easy to observe, and so necessary for the public service of us and our people, that if any person do be remiss hereof we shall esteem all such persons contemptuous and ill affected both to our person and estate, and are resolved to proceed to the punishment of that offender with what severity we may.[5]

There have been many 'England expects' announcements over the years, but this must surely be one of the most remarkable. It was all going too far, and the powers of the saltpetre searchers and gatherers were curtailed by an Act of Parliament in 1656.

An inadequate supply of nitrates was not the only problem with which English saltpetre manufacturers of that era had to contend. Wood ashes were always in short supply because they were required for making soap as well as gunpowder. The clash was resolved by the Lords of the Admiralty directing that the saltpetremen should have priority. Gunpowder was more important than soap!

The home industry was, then, never very successful and it declined, despite many efforts to keep it going, as soon as the East India Company began to import Indian saltpetre on a regular basis around 1630. By 1760 Indian supplies were four times cheaper than the home product.

In France an edict of 1540 granted salpêtriers even greater powers than their

English counterparts, and many of them lasted until 1840. By 1630, 1,600 tons of saltpetre were being produced annually but this had halved by 1775 owing to the competition from Indian imports. To try to boost the home production, the salpêtriers were given even greater powers; the Paris Academie des Sciences offered a prize of 4,000 livres for improved methods of production; and the administration of the Ferme des Poudres, which had become corrupt and inefficient, was replaced by the Régie des Poudres, who appointed four commissioners, including Antoine-Laurent Lavoisier, to be responsible for gunpowder manufacture. M. and Mme Lavoisier went to live in the Arsenal in Paris where they presided over many dazzling social occasions and where he carried out his famous experiments on the nature of combustion.

All this was timely, because France had to rely entirely on home production of gunpowder during the British naval blockade of 1792. Needs must, and in one year more than 7,000 tonnes of saltpetre were produced in about six thousand factories all over the country. Much important research was carried out during the nineteen years in which Lavoisier was associated with the Régie des Poudres, and the quality of French gunpowder improved enormously, a fact which caused much concern in England. Yet in 1794, at the early age of fifty-one, Lavoisier was sent to the guillotine. He had been charged with attempting to supply gunpowder to France's enemies and, in his capacity as a Fermier-général, concerned with the collection of indirect taxes, 'of adding to tobacco, water and other ingredients detrimental to the health of the citizens'.[6] In the general state of turmoil following the start of the Revolution in 1789, a tribunal found Lavoisier guilty and sentenced him to death, saying that 'the Republic has no need of men of science'. But the Republic still needed gunpowder, and Lavoisier's tragic demise was bemoaned by one of his many supporters, Joseph Lagrange, with the words, 'It required but a moment to strike off this head, and probably a hundred years will not suffice to reproduce such another'. In 1840 the last of privileges granted to the salpêtriers were abolished, and by 1870 there were only two factories making gunpowder in the whole of France.

*   *   *

The decline in saltpetre manufacture in Europe and, eventually, in India was accelerated by the discovery of immense deposits of caliche in Chile at the start of the nineteenth century. The caliche, which occurs at a depth of between 0.5 and 4.5 metres in the Atacama region of Northern Chile, contains up to 70 per cent sodium (not potassium) nitrate, and has come to be called Chile saltpetre. The deposits are said to have been discovered

> by a native woodcutter named Negreiros, of the Pampa of Tamataragul, by his having made a fire at a certain spot, which still preserves his name, and observing that the ground thereupon began to melt and run like a stream. He hastily reported this fact to his curé in Camina, who declared it to be hell-fire, and asked to be shown the spot so as to be able to deal with it. The curé took a sample of the salt and found that it was nitre. He threw the remainder into his

garden, where, to his surprise, the plants now grew better than ever before. A British naval officer, visiting Tarapaca some time later, paid a visit to the curé and spread the news to Europe.[7]

For about a hundred years, starting in 1825, when the deposits first began to be mined, caliche played a very important part in Chile's economy, providing almost all the world's supplies of nitrates. These were mainly used as nitrogenous fertilizers and in making nitric acid, with much smaller amounts being used in making gunpowder.

The sodium nitrate in the caliche cannot generally be used directly as a component of gunpowder because, unlike potassium nitrate, it picks up moisture from a damp atmosphere (it is deliquescent). Conversion of the sodium nitrate into potassium nitrate could not be carried out on a large scale so long as the only readily available potassium compounds had to come from wood ashes, kelp, raw sheep's wool or sugar-beet molasses (vinasse), but it became relatively easy when the vast deposits around Stassfurt in Germany, discovered in 1839, were mined for carnallite, which contains potassium chloride. Potassium nitrate made from sodium nitrate and potassium chloride was called conversion or German saltpetre. It first became important in the manufacture of gunpowder during the Crimean War (1854–6) when existing supplies from Europe and India could not meet the demand, and it was widely used thereafter.

*   *   *

Sulphur, commonly called brimstone (burning stone), is a yellow solid which melts at a low temperature. It has been known since ancient times and is mentioned five times in the Bible, for example in Genesis xix 24: 'Then the Lord rained upon Sodom and upon Gomorrah brimstone and fire from the Lord of Heaven.' Homer mentions its use as a fumigant in the *Odyssey*: 'Bring sulphur, old nurse, that cleanses all pollution and bring me fire, that I may purify the house with sulphur',[8] and Pliny refers to its use in medicine, in bleaching cloth, in fumigating wool and in making lamp-wicks for easy kindling.

In early days sulphur was always surrounded by an aura of mystery, and it was regarded as a symbol of evil power, perhaps because of its occurrence in volcanic regions and because of the pungent smell it emits when it burns. Milton, Coleridge and Southey all associated it with hell or with the devil. But all such fantastic ideas are long since discarded, and sulphur was recognized as an element in 1809, since when it has become an essential raw material in the chemical industry.

It occurs in most parts of the world both as natural sulphur and in the form of metallic sulphides, such as fool's gold, and sulphates, like Epsom salts and gypsum. Some natural sulphur is found alongside volcanoes and mineral springs, but the main supplies lie underground, mixed with gypsum and limestone. Until 1900 most of the world's sulphur came from the volcanic regions of Sicily but it now comes from underground sources in Texas and Louisiana in the United States.

Refining sulphur, a sixteenth-century etching. The impure sulphur was heated in a small earthenware pot (A). This boiled the sulphur and the vapour passed into a larger pot (B), where it liquefied. The liquid was then run off, from the bottom, into another container in which it solidified.

In the early Sicilian process, sulphur-containing rocks were laid out in a flat bed on sloping ground and covered with earth and ash in what was called a calcaroni. The rocks at the top of the slope were lit and, as the heat moved downwards, it melted the sulphur in the lower part, which then ran down and could be collected. The process was very wasteful because about 40 per cent of the sulphur was burnt away, and the noxious gases produced devastated the countryside for miles around. But no satisfactory alternative fuels were available in Sicily. The crude sulphur from the calcaroni had to be refined before it was suitable for use in gunpowder, and this was done either on site at some gunpowder mills or by specialist refiners originally concentrated around Marseilles.

Sulphur production was revolutionized in 1894 by Herman Frasch, who had trained as a pharmacist in Germany. After emigrating to the United States, he turned to chemistry so that he could join the fast-growing oil industry, and by 1876, when he was only twenty-five, he had patented a process for removing objectionable sulphur from crude oil. This brought him into contact with the Standard Oil Company of Cleveland, with whom he signed a business agreement, which eventually led to him becoming their first director of research and development. Frasch took out sixty-four patents in his lifetime. Twenty were concerned with refining oil; others with making paint, washing soda, electric light filaments and waxed paper. But it is for his patented method of mining sulphur that he is best remembered.

The sulphur in Texas and Louisiana occurs in thick beds about 150 m or more below the surface. It is mixed with limestone and covered by quicksand, soil and rock, and it had resisted all conventional methods of mining for many years. Frasch patented the idea of using a pump, made up of three concentric pipes of diameters 150 mm, 75 mm and 25 mm, lowered down a hole drilled into the sulphur bed. Hot water, under pressure, was forced down the outermost (150 mm) pipe to melt the sulphur at the bottom of the pump. Hot, compressed air passed down the innermost (25 mm) pipe and forced the molten sulphur up the central (75 mm) pipe as a frothy mixture with water. Sceptics said it would not work, and Frasch wrote: 'Everyone who expressed an opinion seemed to be convinced that this thing could not be done, one prominent man offering to eat every ounce of sulphur I ever pumped.'[9] Later, he described the success of his first test, which ended on Christmas Eve 1894:

I had the heated water run down, under pressure, for twenty-four hours. Then the pumping engine was started on the sulphur line. More and more slowly went the engine, more steam was turned on – until the man at the throttle sang out excitedly, 'She's pumping.' In five minutes the outlet valve was opened. A beautiful stream of golden fluid shot into the barrels we had ready. In about fifteen minutes the forty barrels we had were full. We threw up embankments quickly and lined them with boards to catch the sulphur that was gushing out. . . . When everything had been finished, the sulphur all piled up in one heap, and the men departed, I enjoyed all by myself this demonstration of success. I mounted the sulphur pile and seated myself at the very top.

The process was not immediately profitable because it was so expensive to heat the 20,000 litres of water required for each ton of sulphur, but economic success came within a few years, when it became possible to heat the water by locally discovered oil, and the industry has thrived ever since, contributing to the large number of millionaires in Texas.

*   *   *

Wood charcoal has been used as a fuel and in the extraction of metals from their ores for many centuries so that charcoal burning was well established as a craft long before the advent of gunpowder. Logs of wood were arranged regularly in large conical-shaped mounds with some sort of flue down the centre. All the spaces were filled as fully as possible with smaller pieces of timber to limit the amount of air and the mound was then covered over with turf and rubbish. A smouldering fire was started by forcing burning charcoal or wood down the flue which was then blocked. By making vent-holes around the surface of the mound,

Charcoal burning in the early eighteenth century. Logs of wood from a stack (A) were packed into a conical mound with a flue down the centre (B) and covered with turf and rubbish (C). A smouldering fire, started in the flue (D), converted the wood to charcoal as it spread outwards (E and F). (Reprinted from Fire *by Hazel Rossotti (1993) and* A History of Technology *by C. Singer, E.J. Holmyard, A.R. Hall and T.I. Williams (1954–8), by permission of Oxford University Press)*

the smouldering was made to spread outwards from the centre to convert as much of the wood as possible into charcoal.

Gunpowder was originally made using this sort of charcoal but it became clear, as early as the fourteenth century, that it was necessary both to heat and select the timber much more carefully in order to produce higher quality gunpowder. The precise nature of the charcoal used greatly affects the properties of a gunpowder and choosing the right charcoal was one of the subtle arts of a manufacturer. A much improved charcoal could be made by heating the wood in iron cylinders so that the length of heating and the temperature could be controlled more carefully than in a pile; the product was known as cylinder charcoal and it made better gunpowder. Further improvement came with the correct choice of timber. Roger Bacon's recipe had specified hazelwood, but dogwood, willow and alder were commonly used in England: dogwood for rapid burning, small-grain powder, as used in rifles and shotguns, and willow and alder for coarse-grain powders, which burnt more slowly and were used in larger guns and for blasting. In other countries, poplar, lime, yew, oleander, vine and hemp stems were all used, and beech and birch are satisfactory for cheaper gunpowders.

Wherever possible, the trees were grown close by the gunpowder mill. It was lucky that the amount of timber needed for making gunpowder was not very great because there were many competing demands. Much of the available timber was required for building ships, and charcoal was also needed in large quantities both

Mixing gunpowder in a mortar and pestle. From La Pyrotechnie de Hanzelet Lorrain, *1630. (The Hagley Museum and Library)*

as a fuel and for smelting iron. The resulting deforestation in many parts of the world caused a constant rise in the price of charcoal and led to various Acts by Queen Elizabeth I and other monarchs limiting the felling of trees for iron-smelting.

*   *   *

At first gunpowder was made simply by powdering the three components separately and then grinding them together in a mortar with a hand-operated pestle. Later, stamp mills were used, the mixture being pounded in wooden mortars by wooden-headed stamps which were moved up and down, originally by hand but, from the middle of the fifteenth century, by using horse- or water-power. The method was always hazardous because the friction and percussion from the pestle frequently caused the gunpowder to burn or explode. As a result stamp mills were banned in the United Kingdom by the 1772 Gunpowder Act,

A late medieval stamp mill for making gunpowder. From a South German Firework Book (I.34), mid-fifteenth century. The stamping seems to be being timed by a sandglass, on the left, and it may be the first example of timing an operation in the chemical industry. (The Board of Trustees of the Royal Armouries)

though they were still allowed to be used in making a particularly good sporting powder called Battle powder after the name of one of the places, in Sussex, where it was produced.

Most early gunpowder, which came to be known as serpentine, was not, however, very successful. It was finely, but not evenly, powdered so that it burnt irregularly; it fouled firearms after only a few rounds because it was so dusty and left such a heavy deposit on burning besides producing a lot of nasty smoke; it was greatly affected by damp; the components tended to separate out in transit or storage; and it is now realized that it contained too low a proportion of saltpetre and that what it did contain cannot have been very pure. It was so finely powdered that it was very easy to ram it into a gun barrel too tightly, which resulted in the powder burning far too slowly to be effective. The use of serpentine was, indeed, an art in itself. As William Bourne of Gravesend in Kent, a self-educated mathematician and gunner, wrote in 1587, in his *Art of Shooting in Great Ordnance*: 'The powder rammed too hard . . . it will be long before the piece goes off. . . . The powder too loose . . . will make the shot to come short of the mark. . . . Put up the powder with the rammer head somewhat close, but beat it not too hard.'

Serpentine, with all its defects, was the only readily available powder until

An incorporating or rolling mill with the wheels or runners operated by water-power.
(The Hagley Museum and Library)

about the middle of the sixteenth century, but by that time much progress was being made in producing a more versatile, better quality product. The percentage of saltpetre was steadily increased from the 41.2 per cent of Bacon's recipe to 75 per cent, which became the common figure around 1700 and has remained so ever since. The methods of purifying the components, particularly the saltpetre, were greatly improved, and the replacement of stamp mills by incorporating (or rolling) mills in around 1740, and the introduction of the important process of corning, or granulation, from the middle of the fifteenth century, enabled the blending to be carried out much more effectively.

An incorporating mill consisted of two heavy, wide wheels, known as edge-runners, which were placed parallel to each other a few millimetres above a flat circular bed with a raised edge. The wheels, which were about 2.5 m in diameter and 0.5 m wide, were operated through a system of gears and ran over and around the bed on which was spread a partially mixed gunpowder, moistened with distilled water. The wheels and bed were originally made of stone, as in the old flour-mills, and then of cast iron, but steel was eventually used, with wheels weighing up to 7 tonnes. The pressure of the runners slowly and remorselessly crushed the mixture and ground it together without subjecting it to any very severe shocks, and the milling process was carried out for up to eight hours depending on the product required. The resulting hard mass, called mill-cake or press-cake, could be compacted even further between plates in a hydraulic press, and cake of different densities could be made by altering the duration and intensity of the milling. The three components in the cake were in very intimate contact and bound firmly together so that they would not separate out as in serpentine.

In the corning process, the mill-cake was passed through a series of rollers which broke it down into smaller and smaller grains, the different sizes being separated by sieves with different-sized mesh. Any grains that were too small or too big were reprocessed. The sieved grains were polished by rotating for up to six hours in a drum, and were generally glazed by adding a little graphite during that process. The corning produced grains which were quite strong, and, particularly when glazed, were free-running and resistant to moisture. They could not be packed into a gun barrel as tightly as serpentine so they burnt more rapidly, so much so that 1 kg of corned powder was as effective as 1.5 kg of serpentine.

Corned powder was, however, more expensive and its greater power was, at first, a mixed blessing. It could be used very satisfactorily in small arms, as it was from around 1550, but it was too powerful for early cannon. As Peter Whitehorne wrote in 1560, in his *Certain Waies for the Ordering of Souldiers in Battleray*: 'If serpentine be used in hand guns, it would scant be able to drive their pellets a quoit's cast from their mouths; and if corned powder be used in pieces of ordnance, without great discretion, it would quickly break or mar them.' It was, then, not until the end of the sixteenth century, by which time metallurgy had caught up with chemistry, that large barrels strong enough to withstand the pressures from corned powder became common. And after 1650 the Navy, who suffered particularly from the effect of moisture, used only corned gunpowder.

*   *   *

The introduction of the incorporating mill and the corning process greatly increased the variety of gunpowder products that could be made to meet the demands of many different uses. All the uses were very much alike in kind, but differed greatly in degree. There was, for instance, a vast difference between projecting a bullet from a rifle and a shell from a heavy gun. Similarly, it was clearly much easier to blast sandstone than granite. In each case, however, it was a matter of building up a suitable pressure at a suitable rate and maintaining it for a suitable time. It was, in general, the rate of burning of the gunpowder that mattered most, and this depended on its composition and density, on the shape and size of the grains, and on the surface treatment. The main factor concerned was the total surface area because the powder can only burn inwards from its surface. Thus, for equal weights, a fine-grained, unglazed powder, made of low density, porous material would have a higher surface area and would burn more rapidly than a large-grained, highly glazed, dense powder.

The powders which burnt more rapidly were suitable for use in shotguns, rifles, pistols and revolvers. Because they used relatively light shot or bullets, no very high pressure was needed, but it had to build up rapidly because the barrels were relatively short and the push on the projectile could not be prolonged. It all had to happen quickly, so the powder had to burn away completely in a short time. Larger guns needed a powder which burnt more slowly and for a longer time so that the pressure exerted on the heavy shell could be built up slowly and maintained while the shell was moving along the long barrel.

The main British service gunpowders were classified as RFG (rifle grained fine), with diameters between about 1 mm and 2 mm, and RLG (rifle grained large), with diameters between about 2 mm and 6 mm. On an American scale, which was also used in Britain, the grain sizes were designated as 7F, 6F, 5F and so on down to 2F, F, or C, 2C, and on up to 5C, 6C and 7C as they rose from about 2 mm to 15 mm in diameter. Still larger guns, with bores greater than about 17 cm, required grain sizes even higher than RLG and to meet this requirement different shapes and sizes were cut or moulded from press-cake. Pebble powder consisted of cubes, the P grade with a side of 16 mm and the P^2 of 38 mm. Prism powder had hexagonal-shaped grains.

The burning area of these larger sizes could also be controlled very carefully by drilling holes into them. As a solid cylinder of powder burnt the surface area got less and less so that the rate of burning decreased. As a result the pressure on the projectile in a gun barrel fell away after its first very rapid rise when the gun was fired. If holes were drilled in the cylinder, burning at these inner surfaces increased their area, just as burning at the outer surfaces lowered theirs. By having enough perforations to provide an initial, inner surface area that was equal to, or greater than, the outer one, it was possible to ensure that the pressure on the projectile in a gun rose steadily as the projectile was pushed up the barrel. This gave particularly good results with large guns.

A gunpowder, first introduced by Castner in 1882 and containing a charcoal made by only partially carbonizing rye-straw, was especially good for making perforated pellets. Because of its colour it was known as brown or cocoa gunpowder. Cylindrical pellets with seven central perforations were used by the American Navy in its battle against the Spanish in Manila Bay on 1 May 1898. Five Spanish ships of the line and 1,875 men faced Admiral Dewey's four similar vessels and 1,748 men. The Spanish lost all their ships and suffered 400 casualties, whereas the Americans had only seven men slightly wounded and none killed.

The mining requirements were not as diverse as those of gunnery but a number of different shapes were fabricated. Bobbin powder, for example, was in the shape of cylindrical bobbins with a central hole through which a fuse could be threaded, and similar perforated pellets were made with a triangular cross-section.

The simple mixing of three materials had become a very sophisticated science and art.

* * *

Although the records show that some gunpowder was being stored in the Tower of London as early as 1338, Britain relied on imported powder until the middle of the sixteenth century. This was very unsatisfactory, particularly in time of war; because of the threats from Spain during the reign of Queen Elizabeth I, the government took active steps to encourage home production by granting monopolies and licences. As a result, and because making gunpowder seemed likely to be lucrative, privately owned mills sprang up in many places. Henry and Thomas Lee were making gunpowder at Rotherhithe in 1555, and perhaps as early as 1530; George Evelyn had mills at Long Ditton and Godstone in 1561; and production soon began at Battle and Brede in Sussex, at Faversham, Dartford and Tonbridge in Kent, at Waltham Abbey in Essex, at Hounslow in Middlesex, and at Chilworth and East Molesey in Surrey. The sites chosen were in the southeast of England, close to the main centres of military and naval activity; they were close to a port, or on a river, so that imported nitre and sulphur could be shipped in; they were close to afforested areas so that the wood needed for making charcoal was nearby; and they had good water supplies to operate the water-mills needed for driving the machinery.

It was by no means plain sailing for the early entrepreneurs and there is no evidence that any great fortunes were made. The firms changed hands very frequently, manufacture was never on a very large scale, and it was still necessary to import gunpowder to meet the ever-growing demand. The situation improved somewhat at the start of the seventeenth century when the East India Company began to organize the import of nitre from India and also set up its own powder mills in England, but in the end it became necessary to take the manufacture of gunpowder under government control to ensure supplies of the right quality and quantity. In 1760 the Master-General of the Ordnance arranged to buy the works at Faversham from Thomas Pearse for £5,682; in 1787 the Waltham Abbey Powder Mills were bought from John Walton for £10,000; and a third government plant was established in 1794 at Ballincollig, near Cork in the Republic of Ireland.

These developments were very timely because it became clear during the American War of Independence between 1775 and 1783 that English gunpowder was inferior to that of the French, and a British government inquiry was instituted to examine the situation. This, fortunately, stimulated experimental work at Waltham Abbey which led to such a marked improvement in English gunpowder that it was better than that of the French during the important Napoleonic Wars between 1789 and 1815.

A decline in demand after the wars forced the closure of the works at Cork, while the Faversham works were first leased and then, in 1825, sold to John Hall and Son for £17,935. That firm was destined to play an important role in future developments, and gunpowder and other explosives were manufactured at Faversham under a number of different managements until 1934. The government retained control at Waltham Abbey, renaming the site the Royal Gunpowder Factory. The 6,000 tonnes of gunpowder made there in 1787 rose to 20,000 in 1804 and to 25,000 in 1813, and it also contributed greatly in later years, making a variety of explosives until 1943 when it was closed as a manufacturing

An early advertisement for gunpowder taken from an 1890s edition of the London Post Office Directory. *(The Faversham Society)*

unit. Thereafter, the site was occupied by a number of government agencies, notably the Experimental Research and Development Establishment (ERDE) and the Royal Armament Research and Development Establishment (RARDE), until it was closed in 1991.

The original firms had initially been set up to provide gunpowder in time of war but, as it came to be used in mining during the eighteenth century, it became profitable to set up as a powder maker in the mining areas. To that end, many private mills were established, particularly in Somerset, Wales, the Lake District and Scotland, and by 1800 at least fifty people had tried their hand at making gunpowder. Later, as the centre of mining shifted to the West Country, many businesses were set up in Cornwall, the first one of any real size opening at Cosawes in 1809. They aimed to provide the tin and copper mines with cheaper powder than was obtainable from 'up-country' firms.

By 1875, when making gunpowder came under the jurisdiction of the new Explosives Act, there were twenty-eight licensed manufacturers in the United Kingdom, but that was about the peak of activity. The decline in mining activity and the introduction of better explosives meant that gunpowder manufacture had ceased in Cornwall by 1900. The powder was still required, however, for some blasting in slate quarries and for making fireworks and fuses and there were still sixteen licensed manufacturers in 1920, but by 1939 the only sources of gunpowder in the United Kingdom were plants at Ardeer and Roslin in Scotland. A third wartime factory was opened at Wigtown, also in Scotland, in 1940 but it closed at the end of the war. Roslin continued until 1954 and Ardeer until 1976. Since then gunpowder has been imported into the United Kingdom, mainly from West Germany, although some is still manufactured in the USA.

Many of the operations used in the older processes can still be seen in action at gunpowder museums in England (at the Chart Mills in Faversham); in the Republic of Ireland (at the Ballincollig Royal Gunpowder Factory); in Denmark (at the Frederiksvaerk Museum); in the USA (at the Hagley Museum in Wilmington); and in Tasmania (at the Penny Royal Mills in Launceston). The site of the old Royal Gunpowder Factory at Waltham Abbey is also being redeveloped to include a museum.

CHAPTER 3

The Powder Trust

Most of the gunpowder used in America during the seventeenth and early eighteenth centuries was imported from England, and British colonial policy did not encourage anything that would upset that state of affairs. As a result there were no mills making gunpowder in America just prior to the War of Independence, and the situation looked bleak for Washington's army when the British Parliament prohibited any further exports to America in October 1774. But needs must, and adequate supplies were scraped together by capturing stocks kept in local storehouses, by seizing vessels carrying gunpowder off the coast of Carolina and Georgia, by greatly increasing imports from France and the Netherlands (an operation organized by a secret committee headed by Benjamin Franklin), and by a crash programme of home production. Local small-scale manufacture was encouraged by the issue of pamphlets explaining how gunpowder could be made and by financial grants to suppliers of powder or of sulphur and saltpetre, and each state set about establishing its own mills. Pennsylvania, with a central position and a large German population with some experience of making powder, was the main area of activity and the State Committee for Erecting Powder Mills opened five in 1776.

Supplies of the vital powder were nevertheless scarce and it had to be used with great caution. At the siege of Boston in the summer of 1775, Washington learnt that there were only 90 barrels of powder available instead of the expected 300, and the famous order at Bunker Hill – 'Men, you are all marksmen; don't one of you fire until you see the whites of their eyes'[1] – was issued both to ensure accuracy of aim and to conserve powder. Franklin even suggested issuing pikes, and bows and arrows, to the troops as alternative weapons.

Yet the war was won, the American gunpowder industry had been started, and the new country had realized that its future supplies of gunpowder had better be under its own control. By 1791 the Secretary of the Treasury was able to report that 'no small progress has been made in the manufacture of this very important article. It may, indeed, be considered as already established, but its importance renders its further extension very desirable.'[2] That extension came about mainly with the arrival of the du Pont family from France in 1800.

* * *

Pierre Samuel du Pont, who rose to be Inspector General of Commerce in the cabinet of Louis XVI, was born in 1739, the son of a watchmaker. He was a clever man, never short of grandiose ideas, and with many intellectual and influential friends such as Rousseau and Lavoisier. In 1763 he added 'de Nemours' to his name to indicate his place of residence and to distinguish himself from other du Ponts. They were tumultuous times in France. Lavoisier was beheaded in 1794, and Pierre Samuel, who was regarded as very right wing by the revolutionaries, narrowly escaped the same fate. His opinions were, certainly, strongly held and not to everyone's liking. He was, for instance, a firm follower of Francois Quesnay, a French political-economist who was the leader of the physiocrats. They believed that society should be governed by an inherent natural order, with land and its products as the only source of wealth and the only proper subject for taxation.

It was in this climate that Pierre Samuel set sail for America, with thirteen close relatives, to start a new life. They sailed from Ile de Ré in a dirty, ill-prepared, cramped cargo ship called the *American Eagle*, and, after an appalling crossing of the Atlantic, lasting ninety-one days – some three weeks longer than Columbus took – they landed at Newport, Rhode Island, on New Year's Day 1800. There was a howling blizzard but Pierre Samuel was full of hope, with plans to set up a large company to be involved in speculative land deals close to the frontiers which were just beginning to open up. The family went to live close to New York at a house called Good Stay, which they re-christened Bon Sejours, and he wrote to Thomas Jefferson to announce his arrival.

But neither that, nor anything else, brought any of his schemes to fruition and it was only a lucky break that turned the tide. One of Pierre Samuel's two sons, Eleuthère Irénée, who had left France with him, was interested in science and had worked at the Arsenal in Paris under Lavoisier, at a gunpowder mill in Essones and as the manager of a saltpetre refinery, so he was well versed in the art of making gunpowder. He was also fond of hunting and used to go on expeditions in America with Major A. Louis de Tousard, a former artillery officer in the French army who was now responsible for buying gunpowder and cannon for the American army. One day in 1800, so the story goes, the two men were out hunting using some newly acquired American-made powder in their guns. They were so frustrated by the number of misfires that they realized how poor the American powder was compared with that made in Europe and the major suggested that Irénée should use his knowledge of making gunpowder to set up his own business.

The family was not, at first, very keen on the idea, even though their resources were dwindling. Irénée was only twenty-nine, could not speak English very well and knew nothing about American business methods. But he worked out the details of his new venture so fully that he eventually persuaded everyone that it was viable. With his father's blessing, and with many introductions to old contacts, he returned to France for six months to inspect new gunpowder mills, raise capital and buy machinery. Back in America his friends helped him to find a remote 400 square metre site on Brandywine Creek, about 8 km upstream from Wilmington in the state of Delaware, which was well suited for converting into a powder mill. It was close to a good port; had plenty of water-power, with an existing dam; there were several small buildings and the foundations of a cotton-spinning mill which had been destroyed by fire; and ample supplies of willow

A re-enactment of an 1801 meeting between E.I. du Pont and President Thomas Jefferson to discuss the location of the first American powder mill. (The Hagley Museum and Library)

close by. The site was purchased for $6,740 from the Quaker farmer who owned it.

It took Irénée longer, and cost him more than he had budgeted, to complete the buildings, and some of his French backers and local bankers were worried about their investments until the first powder, under the DuPont label, was produced in April of 1804 and E.I. du Pont de Nemours & Company were in business. It was capitalized at $36,000 with eighteen shares of $2,000 each.

There could not have been a better time to start. Irénée was confident that the methods used by the existing American powder makers were so old-fashioned that they would not provide much competition, and the market was about to boom. Between 1800 and 1900 the population of America grew from 5 to

The DuPont powder yards on Brandywine Creek, 8 km upstream from Wilmington, Delaware, in 1876. (The Hagley Museum and Library)

76 million, and the vast continent was turned from virgin countryside into a nation of towns, roads, canals, rivers, dams and railways scattered with large mining, oil and other industrial enterprises. This expansion alone would have required huge quantities of gunpowder, but the naval and military requirements also increased.

After recent confrontations with both France and England, the United States began to strengthen its defensive power at the turn of the nineteenth century, building up its navy and expanding its army. The Military Academy at West Point, for instance, was established in 1802. All that was well-timed because there were battles to be fought against the British in 1812–15 and against the Mexicans in 1845–8, and there was the Civil War between 1861 and 1865. In 1898 there was the Spanish–American War, the last important conflict in which gunpowder was used in large quantities.

It was not, however, all plain sailing for the DuPont company. The wars caused massive fluctuations in the demand for gunpowder; there were over 200 other manufacturers in the United States by 1810, even though most of them were very

> Dear Sir   Monticello Nov. 4. 11.
>
> I recieved, some time since, the keg of powder, you forwarded for me, and I have been daily expecting you would be so kind as to send on the note of the cost that I might remit it to you. the object of the present is to pray you to forward me another quarter of a hundred, comprehending half a doz. cannisters of shooting powder as before, & the rest proper for blowing rock, of which I have much to do, & to send a note of the cost of both parcels, which shall be immediately remitted you. both qualities have been found of very superior ~~good~~ kind, and having distributed the cannisters among the merchants & gentlemen of this quarter, I presume it will occasion calls on you from them. accept the assurance of my great esteem & respect.
>
> M. Dupont   Th Jefferson

A letter from President Jefferson to E.I. du Pont dated 4 November 1811. Notice the spelling error in the first line. (The Hagley Museum and Library)

small and many of them only operated part-time; and the du Ponts, in building up what was to become a major industry, would suffer over a hundred fatal accidents between 1815 and 1865.

* * *

Eleuthère Irénée ran the company until his death, in 1834, at the age of sixty-three. He left four daughters and three sons, Alfred Victor, Henry and Alexis Irénée. Alfred was then thirty-six and had sixteen years' experience in making gunpowder behind him but he was more interested in the chemical side than in business matters; nor was he in very good health. Henry had hoped for a career in the Army and had entered West Point at the age of sixteen, graduating in 1833. On his father's death, he was recalled into the family fold to serve the company, but he was, at the time, too young and inexperienced to take over. Alexis was only eighteen.

Eleuthère was, therefore, succeeded by his son-in-law Antoine Bidermann, who was married to his daughter Evelina, and had worked with him for twenty years. He reorganized the young company, paid off all its outstanding debts and converted it into a partnership on 1 April 1837, with Eleuthère's seven children as the partners. The company really did belong to the family now and they took the risk of appointing Alfred Victor as President. Alas, events soon proved that he really was no businessman and the company was in debt to the tune of $500,000 by 1850 when the partners' patience ran out and they reluctantly issued an ultimatum which forced Alfred to give way to his younger brother. Henry was only slight in build, but with his upright carriage, red hair, red beard, frock coat and stove-pipe hat, he managed to convey a military air of absolute authority. He was a peppery diehard of the old school, very much averse to change, and thrifty to the point of meanness in many of his personal habits. The invention of the typewriter found him still writing with a quill pen; he continued to use candles, but never more than three at a time, even when natural gas became available; and he would have no ornaments or floor coverings in his spartan office. Smoking expensive Havana cigars and a passionate interest in farming were the only obvious luxuries which he allowed himself. He was not of a scientific bent, but he was a shrewd financier and a good administrator, who revelled in the competitive rough and tumble of industrial activity. The company could not have chosen a better man to lead them.

His particular skills were complemented by those of his nephew, Lammot, one of Alfred Victor's sons, who joined the company at the same time after a mainly scientific education at the University of Pennsylvania. He was nineteen years of age, 1.85 m tall, very strongly built and with light grey hair, a broad chin and cold, grey eyes. His outlook was altogether different from that of his uncle, who was twice his age, because he was a forward-looking optimist, interested in research and technical development, who was anxious to put his ideas to the test. He recognized the sort of problems that tend to hit any family firm after a number of years, and he wrote what could well be a salutary warning for all such organizations: 'We, the present proprietors, inherit in our business all the good done by our fathers as well as those things done badly; and, with the march of improvements, one family in a business is more apt to fall behind the world than to be able to keep in the advance. For while improvements are spread far and wide, the errors and mistakes remain as permanent investments.'[3]

The relationship between Lammot and his uncle, with their different outlooks and the generation gap, was not always very happy but in 1850 they faced a formidable task and it was no time for family squabbles. Indeed, the family had to rally round when, in 1857, Alexis was killed. With two of his sons and some workmen, he was dismantling a powder mill when some sweepings caught fire and exploded. He was blown out on to the road and set on fire but he had the presence of mind to fling himself into the nearby mill-race to quench the flames. He then attempted, very bravely, to climb on to the roof of a powder house that was threatened by flaming debris from the original explosion. A second explosion injured him so severely that he died the following day. A fellow manufacturer said, somewhat fatalistically, that 'a powder mill is like a battlefield. A man drops; that is not an unexpected event; the ranks close up and march on.'[4] The du Ponts did just that.

* * *

The Civil War, which brought about an unprecedented surge in gunpowder sales, provided them with an immediate challenge and opportunity. The shelling of Fort Sumter in Charleston Harbor in the state of South Carolina, on 13 April 1861, brought the differences between the Northern (Union) States, led by Abraham Lincoln, and the Southern (Confederate) ones, under Jefferson Davis, to a head and the war began. General Robert E. Lee commanded the Confederate Army and his Union adversary was General Ulysses S. Grant.

There were, at the time, abut sixty powder mills in the United States, but only two of them were in the south – one in Tennessee and one in South Carolina – and they only employed fifteen men between them. But the Southern States had built up their stocks of powder by buying heavily from the North and, after war was declared, by confiscating any powder on which they could lay their hands. They had, nevertheless, to augment their sources of supply very rapidly if they were to be capable of fighting a prolonged war. This they did by enlarging the Tennessee operation and building a completely new plant at Augusta in Georgia. This was one of the first plants to be operated by steam power and it produced about 1,200 tonnes of powder in three years. The cost of the powder was about $2.4 per kg and although this was much cheaper than the $6.6 per kg that had to be paid for foreign powder it was some three times more expensive than the powder available in the North. The Southern States were, then, at a severe disadvantage regarding both the quantity and the cost of their gunpowder. It meant that their soldiers had to go into battle, for much of the time, carrying less than half the standard issue of ammunition.

Moreover, the Northern gunpowder was of better quality than that available in the South. DuPonts had learnt a great deal by supplying considerable amounts to the British and French during the Crimean War between 1854 and 1856, and in February 1858 Lammot set sail for Europe on a fact-finding tour. He was armed with social advice from his mother – 'Try, my son, not to make mistakes in spelling, or in grammar, when you converse. *Those two*, correctly done, show to others that you are well bred and born. If you go to a *regular dinner* party you must wear white gloves. I hope you will get an umbrella and gum shoes, articles you must have in England's wet clime.'[5] He spent three months visiting European powder mills, including those at Waltham Abbey, and discovering, among other things, that American gunpowder was inferior to the English product except for the fact that it made more noise.

On his return home Lammot was widely acknowledged as the most experienced and skilful gunpowder maker in the United States. He had also developed, in collaboration with Captain Thomas J. Rodman, a special powder called Mammoth powder, which could be used much more satisfactorily than standard powder in the larger guns – up to 50 cm in calibre – which were then

Henry du Pont. (The Hagley Museum and Library)

being built. It had highly compressed grains, as big as golf or tennis balls, and its use enabled the Northerners to out-gun their opponents.

But the North did have two problems with which to contend. They had some difficulty in ensuring adequate supplies of potassium nitrate and there was some concern regarding the security of the DuPont plants in Delaware. Fewer than one in four of the citizens in that state had voted for Abraham Lincoln in the 1860 election, which meant that there were many supporters of the Southern cause. The consequent large-scale dissatisfaction when Delaware did not secede meant that the du Ponts had to live with the constant threat that their gunpowder plants might be attacked by internal saboteurs or by raiding parties from the nearby Southern States.

Self-defence was the order of the day and the du Ponts raised two militia companies from their workforce. Henry du Pont, with his military experience, was appointed as commander of all the troops in Delaware. He was given the rank of major-general, and was henceforth known as 'The General'. There was particular concern when General Lee's forces moved into Pennsylvania, which bordered Delaware, but they were driven back at Gettysburg, and a few days later General Grant captured Vicksburg in Mississippi. These two defeats for the Southern armies signalled the end of the war but it was not until almost two years later that Lee surrendered to Grant at Appomattox in Virginia. Lincoln was assassinated four days later. DuPonts probably made a profit of over a million dollars from the war.

* * *

Things were very different after the war when the whole gunpowder industry fell on very bad times. The vast quantities of powder still in store had to be sold off and some of it remained on the market until 1890, being offered for as little as 11 cents per kg. There were far too many manufacturers chasing far too few customers and that led to cut-throat competition, together with bribery and corruption, in a desperate bid for survival. In addition, the gunpowder manufacturers were having to face up to the arrival of new high explosives in their markets.

To try to save their own skins, the three largest manufacturers – the DuPont Company in Delaware, the Hazard Powder Company in Connecticut, and the Laflin & Rand Company in New York State – joined together with four smaller companies to form the Gunpowder Trade Association of the United States, which became known as the Powder Trust. Of the 48 votes involved, the big three had 10 each and the other five had 18 between them. It did not work against DuPont's interests that Lammot was elected as the first President of the Trust. There was nothing illegal about forming a Trust in 1872 and the members would argue that it was essential to bring some sanity to a chaotic situation, in which half the country's gunpowder-making machinery soon lay idle. But it was not a happy event for the non-members, nor, in the end, for most of the members.

DuPonts were certainly determined to hold their own. They slashed the price of their powders to squeeze the weaker brethren out of business and ruthlessly set about taking over the remainder. In 1876 they bought the Hazard Powder Company and by 1889, when Henry du Pont died, he had been so successful in stifling competition and creating a monopoly that the Powder Trust was effectively controlled by DuPonts and Laflin & Rand.

Henry had worked for his company for fifty-five years and been in charge for thirty-nine of them. He was a man of great stature, reflected in his lifetime by his nickname of 'The General', and was commemorated on his tombstone by the verse from the 37th Psalm: 'Mark the perfect man and behold the upright: for the end of that man is peace.' But it is ironic that someone who disliked change so intensely presided over so much of it.

* * *

One change that 'The General' did not see, and one which he would certainly not have liked, was the passing of the Sherman Anti-Trust Act, in 1890, 'to protect trade and commerce against unlawful restraints and monopolies'. This new legislation made no immediate impact and Eugene du Pont, a son of Alexis, who had been working alongside his father when he lost his life, succeeded Henry and continued his policy. He set about modernizing the old-fashioned state of affairs he had inherited from his uncle; he gathered still more companies under the DuPont umbrella; he saw to it that three new gunpowder manufacturers did not survive more than a few months; he negotiated the London, or Jamesburg, agreement which was designed to keep European firms out of the Western hemisphere, and vice versa; he geared up DuPonts to see the country through the Spanish-American War of 1898, making around $500,000 profit in the process;

and he changed the status of the company from a partnership into a corporation in an effort to make it easier to administer now that it had grown so much.

Eugene's death in 1902, a week after contracting pneumonia, caused a family crisis because there was no obvious heir apparent. The clan had grown so much that there was no shortage of candidates, but of those with the du Pont name who came immediately to mind some were too old, some were ill and some were regarded as unreliable. The presidency was thus offered to Hamilton M. Barksdale, who had married into the du Pont family and who worked for the company, but Barksdale thought that the head should be a genuine du Pont, as it had been since the company was founded, save for Antoine Bidermann's short rule in 1834.

So serious was the dilemma that some of the family even suggested selling out to Laflin & Rand for a figure of twelve million dollars, a suggestion that would have had many of the older generation turning in their graves. Nor did it please Alfred I. du Pont, one of the contenders for the presidency, who claimed that the business was his birthright. He turned for help to two first cousins. One was T. Coleman du Pont, who, at the age of thirty-nine, had recently 'retired' from a very successful industrial career outside the explosives field, to devote his energies to farming. The other was Pierre S. du Pont, the eldest son of Lammot.

After much negotiation, this trio – all great-grandsons of the company's founder and all in their thirties – found a way of matching the twelve million dollar figure to take control of the company themselves. Coleman became President; Alfred was Vice-president; and Pierre was Treasurer. DuPont's own plants and property represented around 40 per cent of the company's assets; the other 60 per cent related to holdings in other companies, often shared with Laflin & Rand. Ownership and valuation were, however, so confused that Pierre wrote 'We have no real idea what we are buying.'[6]

* * *

Coleman, nicknamed Coly, had studied engineering at the Massachusetts Institute of Technology and had worked in coal-mining, in the steel industry and in electric street-railways. He was, in many ways, larger than life. Well over 1.8 m tall and weighing more than 89 kg, he had been a formidable boxer and footballer; he was a dynamic extrovert, who liked nothing better than consorting with stars; he was something of a playboy who revelled in practical joking; yet he was also a hard-working, ruthless businessman, who put all his experience, enthusiasm and energy into moving DuPonts into the twentieth century. What a start he made. Within a few months of taking office he had completely revolutionized the situation by buying Laflin & Rand for four million dollars, which meant that the DuPont company and the Powder Trust were now one and the same thing.

With this coup, Coly gained control of over fifty companies spread across the country, and he soon began to knock them into shape. Some were closed down, some sold, some amalgamated, some reorganized. Everything was up for change. The DuPont headquarters moved from Wilmington on to a complete floor of a New York hotel; an Executive Committee was set up to make important decisions; new research laboratories were built; a pension scheme was started; bonuses of stock were given to successful employees; price-cutting, no longer necessary, was abandoned; and new safety regulations were introduced. Most importantly for the future, Coly established a Development Department to investigate ways of diversifying away from the cyclical explosives industry towards general chemical manufacture.

While all this was going on, public and governmental suspicion of Trusts, which were spawning in many industries all over the world, was growing and there were increasing demands that something should be done about them. The President of the United States, Theodore Roosevelt, agreed and an era of Trust-breaking began.

On 7 July 1907 the United States Department of Justice instituted a suit against the DuPont company on the grounds that it had acted to stifle competition in the explosives industry, which was illegal under the Sherman Act. The government claimed that the DuPont company controlled an unacceptably high proportion of the market for explosives and that this position had been reached by illegal methods. The litigation lasted for four years, and built up enormous piles of paper, before the court finally found against DuPonts on 12 June 1911. But, given that the DuPont company was, to a large extent, the explosives industry of the United States, what could be done? The court threw the question back at the government and the du Ponts, directing that the two parties should, between them, devise a plan that the court would find acceptable. So, on 15 December 1912 almost two-thirds of DuPont's business was transferred to two new companies – the Hercules Powder Company and the Atlas Powder Company. Even so, the truncated DuPont company was still worth six times more than Alfred, Coleman and Pierre had paid for it ten years earlier.

*   *   *

The enforced reorganization was a severe setback to DuPonts, but it was only temporary and it did not prevent the company playing a very important role as suppliers of explosives to both the Allied and American armies during the First World War. Coly had seen the company through the anti-Trust legislation battle and had geared it up for the war effort, but he had not been at all well for much of that time and was anxious to lay down his responsibilities. This was achieved in 1915 when he handed over control to Pierre, but only after a long period of family feuding and financial litigation. In 1907 it had been the United States v. DuPont. Now it was du Pont v. du Pont.

It was not a happy period for the family, but when it was all settled, Coly found a new lease of life and began yet another career. He built hotels and office-blocks in New York and elsewhere; he bought up restaurants and insurance companies; he created a modern road system in Delaware, often at his own expense, which earned him the title 'father of the superhighway'; and he became a Senator in 1921. He died, of cancer of the throat, on 11 November – Armistice Day – 1930.

Many of the presidential duties had fallen on Pierre during Coly's illness so he was no stranger to the post when he became President in 1915. The war effort was the main priority at that time, but Pierre had the foresight to prepare for the future. He directed that the Development Department should study ways and means of diversifying away from explosives, and thousands of possibilities were investigated. As early as 1910 the Fabrikoid Company, makers of a new waterproof cotton cloth, had been bought, and the Arlington Company, making lacquers, plastics and enamels, was taken over in 1915. By the end of the war, another seven businesses had been swallowed up.

This set the pattern for the future and the profits made during the First World War provided the funds. DuPonts had flourished in the past by taking over other manufacturers of explosives. In 1913 97 per cent of their business was in explosives and they had assets worth $75,000,000. They were now to build themselves up, predominantly in the general chemical and related industries, but also by buying a large stake in the General Motors Corporation. By 1939 only 10 per cent of the business involved explosives, yet the assets had risen to $850,000,000. The Second World War saw the company once again as a major manufacturer of explosives, also playing an important role in the development of the atom bomb, but after the war the move away from explosives accelerated.

In 1981 the biggest merger in American history saw DuPont's acquire 50 per cent of the stock of CONOCO – the Continental Oil Company of Stanford, Connecticut – creating an organization, sometimes known as 'the new DuPont', with combined assets of over eighteen billion dollars. Today, operating in forty countries worldwide at 200 different sites, the group employs about 110,000 people and has annual sales of over $40 billion.

Tall oaks from little acorns grow.

CHAPTER 4

Testing Gunpowder

Once different types of gunpowder became available it was necessary to devise tests by which one could be compared with another, and these were to have important consequences far beyond the realm of explosives.

Initially, the tests were carried out, very inefficiently, by feel, by burning to see what residue was left, or by burning on a piece of paper and examining the nature of charring. One early recipe[1] for a test shows how haphazard it must have been:

> Lay two or three small heaps (a dram or two) on separate pieces of clean writing-paper; fire one of them by a red-hot iron wire; if the flame ascends *quickly*, with a good report, leaving the paper free from white specks, and does not burn it into holes; and if the sparks fly off, setting fire to the adjacent heaps, the goodness of the ingredients and the proper manufacture of the powder may be safely inferred; but if otherwise, it is either badly made or the ingredients impure.

The oldest known instrument, which became known as a trier or tester, or by the French term éprouvette, is that described by William Bourne in his *Inventions and Devises* published in 1578. It was a small, metal cylinder with a heavy, hinged lid. On firing a sample of gunpowder within the cylinder the lid was pushed open, whereupon it caught on a tooth of a quadrant-shaped ratchet. The angle through which the lid was moved by a certain weight of gunpowder gave a rough indication of the strength of the powder. Robert Hooke demonstrated an improved version to the Royal Society in 1663.

Meanwhile, Joseph Furtenburg had modified the system in 1627, by making the gunpowder in the cylinder raise a heavy, conical-shaped cap which was guided vertically upwards by two wires. The cap was prevented from falling back by a series of catches, and the device was known as a flying-cap trier. Other methods adopted included the expulsion of water from one vessel into another, the firing of a bullet from a pistol into clay, the bending of springs, or the measurement of the distance a ball could be propelled from a mortar.

It was, however, the Trauzl test, invented by Isidor Trauzl, the manager of Nobel's Austrian factory at Zamky near Prague, that became the most popular. It involved measuring the deformation caused by the detonation of 10 grams of an explosive in a hole within a cylindrical block of lead. The conditions under which the test should be carried out were laid down at the 5th International Congress of Applied Chemistry in Berlin in 1903.

The Trauzl test. Right: *A standard Trauzl block, 200 mm long and 200 mm in diameter, with a central hole 115 mm deep and 25 mm in diameter.* Left: *The deformation caused by an explosive. (Bureau of Mines Bulletin, no. 15)*

The Trauzl test is still in use in some countries, but in England and to a large extent in the United States it was replaced in around 1900 by the ballistic pendulum test, which had been invented by Benjamin Robins as early as 1742 for testing guns. Robins was the son of poor Quaker parents and was born in Bath in 1707. It was not easy for him to get much formal education but he was befriended by Pemberton, a friend of Newton's, and this encouraged his interest in mathematics. By the age of eighteen he was teaching that subject in London, and also pursuing his studies in navigation, architecture, fortification and gunnery. He described the pendulum in his *New Principles of Gunnery*, published in 1742. It was one of the earliest attempts to put gunnery on a sound scientific basis. He became a Fellow of the Royal Society and was the Engineer-General to the East India Company before dying in Pondicherry, in India, in 1751.

A modern ballistic pendulum consists of a 5,000 kg mass of steel suspended from a framework so that it can swing freely. The explosive under test is fired in a cannon placed 50 mm away from the face of the pendulum, and the swing of the pendulum is measured on a vernier-type slide rule. That swing is compared with the 83 mm given by a particular gelignite mixture which has been adopted as a standard. Mining explosives give swings between 38 and 70 mm.

In a ballistic mortar, which is used as an alternative to the pendulum, the explosive under test, embedded in the head of a freely swinging pendulum, is used to fire a steel shot out of the head. The recoil of the pendulum is measured and compared with that given by a chosen standard of blasting gelatine.

*   *   *

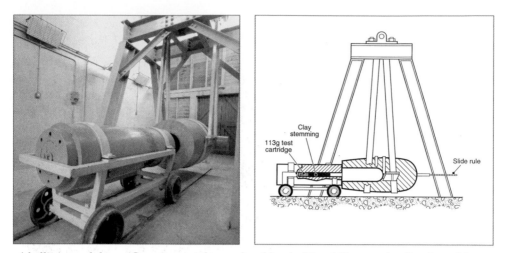

Clay
stemming

113g test
cartridge

Slide rule

A ballistic pendulum. (Crown copyright; produced by the Visual Presentation Services of the Health and Safety Laboratory)

Shortly after the death of Benjamin Robins, Count Rumford continued the same line of research in a series of experiments he carried out between 1778 and 1797. He was a tall, handsome man of fashion, with an exceptional ability, a particularly charming manner, and a very strong social conscience, who had a quite remarkable career. He was born Benjamin Thompson in 1753, into a family of English origin living in North America. After a very limited education he took to teaching and, when he was nineteen, he married the rich, widowed, 32-year-old daughter of his headmaster. This enabled him to settle down to the life of a country gentleman and to pursue his interest in gunpowder. At the start of the American War of Independence, it also enabled him to gain a commission as a major in the American army. But his over-rapid promotion, and his rather obvious display of sympathy for the English cause, led to so much local antagonism that he was forced to flee, eventually to England, leaving his wife and young daughter behind. He never saw his wife again, separating from her in 1775 when she remarried, and he had to wait twenty years before seeing his daughter.

He carried with him to England an introduction to Lord George Germain, the Secretary of State for the Colonies, who offered him a job in the Colonial Office and provided him with accommodation in his country house, Stoneland Lodge in Sussex. It was there that Thompson did a lot of his research on gunpowder. His first scientific paper, published in 1791 under the title *New Experiments upon Gunpowder, with occasional Observations and practical Inferences*, described the results he obtained when he fired a small, freely suspended cannon into a ballistic pendulum under different conditions. He measured both the movement of the pendulum and the recoil of the cannon, and one of his suggestions, not widely taken up, was that it was better to test a gunpowder by measuring the latter rather than the former. He also disproved the belief, widely held among gunners, that

Count Rumford at the age of forty-nine.

damp gunpowder was more effective than dry; he played a large part in the decision to introduce bayonets into the British Army; he tried, unsuccessfully, to 'shoot flame instead of bullets'; and he made a significant number of practical suggestions for improving the design of field artillery.

His interest in gunpowder remained with him throughout his life. He had almost blinded himself as a teenager when he tried to make some fireworks and he later summed up his views when he wrote: 'The force of gunpowder is so great, and its effects so sudden and so terrible, that, notwithstanding all the precautions possible, there is ever a considerable degree of danger attending the management of it, as I have more than once found to my cost.'[2]

His reputation as a scientist led to his election as a Fellow of the Royal Society, and his knowledge of American affairs to his appointment, in 1780, as Under-Secretary of State for the Colonies. He held that office for just over a year before returning to America to fight on the English side. When peace was declared, he returned to England where he was knighted by George III.

Shortly afterwards, on a journey to Strasbourg, he was introduced to the Elector of Bavaria and, as a result, he began a completely new career in that country, where he became Minister of War and of Police. He came to be so highly respected in Bavaria for his administrative reforms, his contributions to industry, his scientific experiments on the mechanical nature of heat, and his practical inventions of heating, lighting and cooking equipment, that he was made a Count of the Holy Roman Empire, taking the title Count Rumford, the old colonial name for Concord, where his wife and daughter still lived. When he left Bavaria in 1795, a memorial was erected in the 'English Garden' which he had laid out in Munich.

Back in England, he continued his very practical social and scientific activities. He started a 'Society for Bettering the Conditions of the Poor' in 1796, and founded the Royal Institution in 1800. This was to be concerned with 'diffusing the knowledge and facilitating the general and speedy introduction of new and useful mechanical inventions and improvements; and also for teaching, by regular courses of philosophical lectures and experiments, the application of those discoveries in science to the improvement of arts and manufactures, and in facilitating the means of procuring the comforts and conveniences of life'.[3]

Rumford acted as the first secretary of the Institution but, disappointed by the lack of public support, he left in 1803 to settle in Paris, where he married Lavoisier's widow. The marriage lasted only four years and, after they had separated, Rumford went to live a solitary life in the suburbs of Paris, looked after by his daughter. He died on 21 August 1814. Few people have done so much, so successfully, in so many different ways. One eulogist concluded with the words that Rumford had 'by the happy choice of his subjects as well as by his works earned for himself the esteem of the wise and the gratitude of the unfortunate'.[4]

*   *   *

The early gunpowder triers demonstrated that the explosive power of gunpowder could be used to lift weights, move water, bend springs, and do work in other controlled ways. The idea that gunpowder might be used for constructive purposes began to dawn. As so often seems to have happened, that idea had already occurred to Leonardo da Vinci for he described and drew the first embryonic gunpowder engine as early as 1508. Yet it was not until 1673 that Christiaan Huygens, a distinguished Dutch scientist, best known for his invention of the pendulum clock and as the initiator of the wave theory of light, described an engine operating on a 'new motive power by means of gunpowder and the pressure of the air'.[5] He wrote: 'The force of cannon powder has served hitherto only for very violent effects, such as mining and blasting of rocks, and although people have long hoped that one could moderate this great speed and impetuosity to apply it to other uses, no one, so far as I know, has succeeded in this; at any rate no notice of such an invention has appeared.'[6]

The Huygens engine was built by his assistant, Denis Papin, at the Academie Royale des Sciences in Paris. It consisted of a vertical cylinder containing a piston. Some gunpowder at the base of the cylinder was ignited to push the piston up, and most of the hot gases were allowed to escape through valves at the top of the cylinder. As the cylinder and the remaining gases cooled, a partial vacuum formed in the cylinder and atmospheric pressure drove the piston downwards. The piston could be connected by a rope over a pulley to lift weights or operate a water pump. It was hoped that it might be possible to pump water out of the River Seine to operate Louis XIV's fountains. The engines were more fanciful

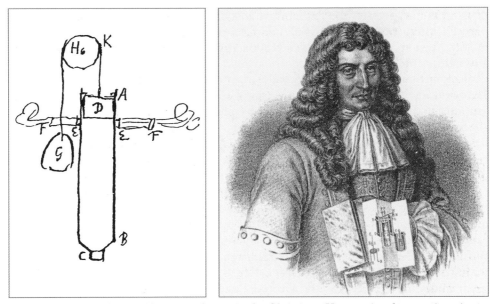

Left: *A drawing of a possible gunpowder engine by Christiaan Huygens in a letter written in 1673*. Right: *Denis Papin, who built a model of the engine.*

than practical, even though Huygens prophesied that they would come to work mills and drive vehicles on land and sea. It was not possible to remove all the gases from the cylinder after the gunpowder explosion so that only a poor vacuum was created, and re-charging the cylinder with powder after every single stroke was both slow and somewhat hazardous.

Papin became a professor at Marburg where he constructed an improved version of the engine in 1688, but he still could not solve the problems. The idea did, however, bear fruit because he eventually realized that, rather than using exploding gunpowder, the piston could be driven in the cylinder by making use of the 1,300-fold expansion of water when it turns into steam. Legend has it that the inspiration came to him as he watched the steam from a boiling kettle lift the lid. He wrote:

> Since it is a property of water that a small quantity of it, turned into vapour by heat, has an elastic force like that of air, but upon cold supervening is again resolved into water, so that no trace of the elastic force remains, I readily conclude that machines could be constructed wherein water, by the help of no very intense heat, and at little cost, could produce that perfect vacuum which could by no means be obtained by the aid of gunpowder. [7]

The unsuccessful gunpowder engine had thus fathered the enormously powerful steam engine. Papin made a number of working engines and, on the side, developed his digester, the forerunner of today's pressure cooker. The seeds he planted grew into the improved steam engines of Savery (1698), Newcomen (1712), Watt (1776) and Trevithick (1800). In turn, the steam engine led on to the gas engine, and the internal combustion engine followed. The use of a gas explosion to move a piston in a cylinder was first attempted by a French inventor, Etienne Lenoir. In 1859 he built a working engine in which a mixture of coal gas and air in the cylinder was exploded by an electric spark, but the first fully successful gas engine was built by a German, Nicholas August Otto, in 1876. The first gas engines had a maximum of about 3 horsepower but by 1917, when they reached the zenith of their popularity, engines of 5,000 horsepower were being used, and they were fully competitive with steam engines.

Towards the end of the nineteenth century, however, both the gas and the steam engines were beginning to be superseded by oil engines operating on much the same principle but using diesel oil or petrol as the fuel. The change came about following the discovery of oil wells, particularly in Pennsylvania, in the United States, around 1860. The black, sticky crude oil – similar to that which had played its part in Greek fire – was originally distilled to obtain paraffin (called kerosene in the United States) which was required for heating and lighting purposes. In those days, the petrol (gasoline in the United States), which was also obtained from the crude oil, seemed to be too dangerously inflammable to be of any use; much of it was simply destroyed by burning. That changed as the oil and petrol engines designed by Rudolf Diesel, Gottlieb Daimler and Karl Benz came into use, and grew into today's internal combustion engines.

They can all be regarded as descendants of the simple gunpowder triers. It is in that context that Professor J.D. Bernal wrote in 1948:

Ultimately, however, it was the effects of gunpowder on science rather than on warfare that were to have the greatest influence in bringing about the Machine Age. Gunpowder and the cannon not only blew up the medieval world economically and politically; they were major forces in destroying its system of ideas. As John Mayow put it – 'Nitre has made as much noise in philosophy as it has in war' . . . The force of the explosion itself, and the expulsion of the ball from the barrel of the cannon was a powerful indication of the possibility of making practical use of natural forces, particularly of fire, and was the inspiration behind the development of the steam-engine.[8]

Shakespeare, unaware of the later developments, had given a different verdict on nitre when, in *Henry IV*, Hotspur says:

> And that it was great pity, so it was,
> This villainous saltpetre should be digg'd
> Out of the bowels of the harmless earth,
> Which many a good tall fellow had destroy'd
> So cowardly: and but for these vile guns,
> He could himself have been a soldier.

CHAPTER 5
'Crakys of War'

The nursery rhyme runs:

> Please to remember
> The fifth of November
> Gunpowder treason and plot.
> We know no reason
> Why gunpowder treason
> Should ever be forgot.

But what is to be remembered of this seventeenth-century whodunnit? The popular version, encouraged by the government of the day and by some contemporary historians, and now well-established folklore, has it that a desperate band of persecuted Roman Catholics conspired to blow up the sovereign lord King James I together with his queen, the prince, all the lords spiritual and temporal, and the Commons, while they were all assembled in the House of Lords for the opening of Parliament, originally intended for 7 February 1605.

The main conspirator was Robert Catesby, a Roman Catholic country gentleman from Warwickshire. 'In that place', he said, 'they have done us all the mischief and perchance God hath designed that place for their punishment.'[1] Catesby recruited several others to his cause: his cousin and friend, Thomas Winter; Thomas Percy, a cousin of the Earl of Northumberland, together with his brother-in-law, John Wright; and a pious, brave, bearded soldier called Guy Fawkes, who was serving as an exile in the Low Countries.

In 1605 the chamber of the House of Lords was on the second floor above some old kitchens which were being used as a store-room by a coal-merchant. The original plot was that a pile of gunpowder would be placed under the coal-store and to that end Percy leased a nearby house, from which the five conspirators began digging a tunnel on 11 December 1604. By Christmas Eve they had reached the foundations of the wall of the House of Lords, when, for no very adequate reason, the opening of Parliament was postponed until 3 October, and then to 5 November. This was lucky for Catesby and his men for they found it difficult to make quick progress in breaching the 3.5 m thick foundations. Their luck held, too, for at this moment, the coal-merchant sold off his stock and gave up the lease of the store-room, whereupon Percy hired it for his own nefarious purposes.

Guy Fawkes with some of his associates. (Adam Hart-Davies)

Shortly thereafter one and a half lasts, that is thirty-six barrels or 1,630 kg, of gunpowder were slowly transferred from Catesby's lodgings across the river in Lambeth into the storeroom, which is now known as Guy Fawkes's cellar. The pile was covered by firewood to hide it. During the summer months, Fawkes made frequent visits to the cellar to see that the stores had not been tampered with, and Catesby widened and developed the plot, taking several others, notably Sir Evered Digby and Francis Tresham, into his confidence. Following the explosion planned for 5 November, Catesby was to proclaim a new, more tolerant sovereign; there was to be an uprising in the Midlands, organized by Digby around a hunting party at Dunchurch; and a force of exiles from Flanders was to land on the south coast.

It was all to no avail. As Lord Monteagle, who was married to Tresham's sister, sat down to supper on Saturday 26 October, one of his servants delivered a mysterious, anonymous letter which had been given to him by a complete stranger. It was clearly intended to warn the noble lord of impending danger. 'This Parliament', it said, 'shall receive a terrible blow, and yet they shall not see who hurts them.'[2] Monteagle immediately raised the alarm, riding off with the

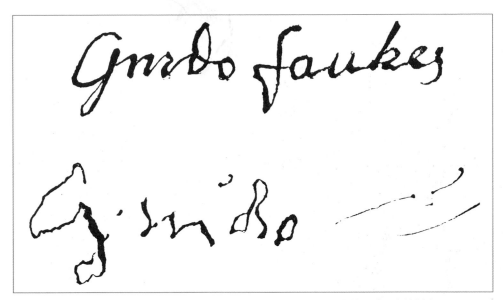

Guy Fawkes's signature before and after torture. (Public Record Office, SP 14/216)

letter to the Earl of Salisbury, King James's principal private secretary, who for the previous nine years had been the power behind the throne. For this not very brave act, Monteagle was proclaimed by Ben Jonson as 'the saver of my country', and was rewarded by the government with a pension of £500 for life and £200 in perpetuity in fee-farm rents.

Guy Fawkes was arrested early in the morning of 5 November in the famous cellar. It is widely believed that he was holding a slow match, made by dipping cords of hemp into a solution of potassium nitrate and drying, which was already smouldering and which had only a further fifteen minutes to burn. Whether that be true or not, he was imprisoned in the Tower in the cell next to the torture chamber, the King having interrogated him and given orders for slow torture to the limit. He confessed fully after three days on the rack. On 31 January 1606 he was executed, quartered and castrated, together with Winter, at an impressively stage-managed spectacle in the old Palace Yard at Westminster. Digby and Tresham met similar fates, but Catesby, Percy and Wright, who had fled to Holbeach House, just north of Bromsgrove, had been killed when a posse led by the High Sheriff of Worcestershire attacked the house on 8 November 1605.

But was it all so? The story rests heavily on contemporary governmental reports, on a confession by Winter, which was probably a forgery, and on the tale squeezed out of Fawkes. Many historians now think that there is at least something in the suggestion that the whole episode was more a conspiracy of government against Catholics than the other way round. Salisbury was a machiavellian figure, who would stoop to almost any depths to discredit Roman

Catholics, and who had his own well-organized secret service. It seems improbable that he, and others, did not know at an early stage what was going on. How could so few amateurs secretly dig a tunnel in that busy part of London? What did they do with all the debris? How did they come by such a large amount of gunpowder, when all supplies had been under the control of the government since 1601? Why are the gunpowder records of the time missing? Why was the opening of Parliament twice postponed? Who wrote the letter to Lord Monteagle? Was it Tresham or one of his friends? Or was it all a plant organized by Salisbury or one of his agents? These questions have never been fully answered. The tales have been told and each must make his choice.

What was the impact? Catholics suffered even more than before; Parliament decreed that 5 November be celebrated every year as a day of public thanksgiving; Guy Fawkes became a national figure and an eternal benefactor to fireworks manufacturers; and gunpowder, described in the Act of Parliament as 'an invention so inhuman, barbarous, and cruel, as the like was never heard of before'[3] got the sort of free coverage that advertising agencies dream of.

* * *

But gunpowder cannot be all bad when it has provided such innocent fun and pleasure in fireworks for so many years. They began around the sixth century, with the Chinese making their first crackers simply by throwing lengths of bamboo cane on to a fire. The expansion of the gases inside the cane eventually burst open the outer casing with a sharp explosion like the firing of a gun. Marco Polo described the effect: 'they burn with such a dreadful noise that it can be heard 10 miles at night, and anyone not used to it could easily get into a swoon or even die. Hence the ears are stopped with cotton wool and the clothes drawn over the head, and horses are fettered on all four feet and their ears and eyes covered. For it is the most terrible thing in the world to hear for the first time.'[4]

Once gunpowder was discovered, some three hundred years later, it did not take the Chinese long to use it in improving their crackers and in making a whole range of other fireworks – rockets, coloured flames and sparklers – which were being described in the West by the thirteenth century. Roger Bacon wrote: 'We can produce in air sounds loud as thunder and flashes bright as lightning',[5] and Marcus Graecus gave some typical recipes in his *Book of Fires*:

> The second kind of flying fire is made in this way. Take 1 lb [454 g] of native sulphur, 2 lb [908 g] of linden or willow charcoal, 6 lb [2724 g] of native saltpetre, which three things are very finely powdered on a marble slab. Then put as much powder as is desired into a case to make flying fire or thunder. Note: the case for the flying fire should be narrow and long and filled with well-pressed powder. The case for making thunder should be short and thick and half filled with the said powder and at each end strongly bound with iron wire.[6]

The essential features of these early fireworks have remained largely unchanged except for detailed refinement over the centuries, but the original association with religious and superstitious rites, such as the scaring away of evil spirits, has been lost in the spectacular effects that can be created in large, ceremonial, triumphant displays, in frolics and fiestas, and in the development of fun fireworks into military rockets and guns.

The simple fireworks rocket consists of a tube, generally made of cardboard, closed at its upper end. The tube is filled with a pressed gunpowder mixture through the open, lower end of the tube and a conical hole is drilled down the centre of the powder. A slow-burning fuse leads into the powder through the lower hole, which is constricted to some extent once the tube is filled. A wooden stick is attached to the rocket so that it can be stood in an upright position for launching. On lighting the fuse, the gunpowder burns from the surface of the conical hole outwards, and the thrust of the hot gases through the constricted hole drives the rocket into the air. The stick helps to maintain the direction of the rocket's flight.

In a display rocket, the lower two-thirds of the tube is filled with gunpowder as in the simple rocket. Above it, and separated by a fuse, there is another compartment containing a small charge of gunpowder and an assembly of small fireworks such as star shells, sparklers, coloured lights and crackers. The fuse within the tube is timed to set off the small gunpowder charge when the rocket is at its highest point so as to discharge the assembly of fireworks across the sky.

Scintillation effects, as in sparklers and Catherine wheels, are obtained by adding metal filings to a gunpowder mix. When the powder ignites it raises the temperature of the filings to the point of incandescence so that they sparkle. Iron filings sparkle with a white or yellow colour; aluminium or magnesium with a white colour; copper with green; and zinc with bluish-white. Coloured flames, as in Roman candles or Bengal lights, can also be achieved by incorporating different chemicals into a gunpowder mixture. In the early days a wide variety of plant or animal products were used to get different colours but later metallic salts were used. Antimony sulphide gives a strong white colour, strontium salts burn crimson, calcium salts red, barium salts green, and sodium salts yellow. Intermediate colours can be obtained by using a mixture of salts, and salts can also be used with powdered metals to give both colour and sparkle.

*   *   *

During the second half of the twelfth century the Chinese appear to have had two sorts of firework rocket, probably made from hollowed-out bamboo canes filled with gunpowder. One was set off in a horizontal position on the ground, whereupon it scurried around in all directions; hence its nickname 'ground-rat'. The other had a stick attached and was launched upwards; it flew off into the air and was called a meteor. The close relationship between the two is emphasized by the later use of names such as 'flying-rat' and 'meteoric ground-rat'.[7] They must have been frightening to more than evil spirits and it is not surprising that they soon developed into military weapons, particularly in a civilization accustomed to hurling incendiary mixtures or shooting fire-arrows at an enemy. By 1312 some sort of rocket was used by the Chinese against the Tartars, and rockets were used

A fireworks display in Cherbourg when the British Channel Squadron visited to celebrate Emperor Napoleon's jour de fête *in 1865. (From* The Illustrated London News, *26 August 1856)*

at the battle of Delhi in India in 1399. In the West their first appearance was at the battle of Chioggia between Venice and Genoa in 1380.

Any damage done by these early rockets must have been more psychological than physical for they can have been only very puny weapons, and they were only used sporadically until they were revived, towards the end of the eighteenth century, in India. In battles at Seringapatam in 1792 and 1799 Hyder Ali, Prince of Mysore, and later Tippu Sultan, his son, used rockets very effectively against the British. Their rockets had a tube 20 cm long and 3.8 cm in diameter, weighed about 5.5 kg, were stabilized by a 3 m bamboo pole, and fired either incendiary or explosive missiles. The rockets had a range of about 1,000 m, could kill three or four men, and had a very alarming effect on horses.

This new Indian method of attack inflicted more casualties on the British than conventional gunfire and convinced them that they ought to investigate it with a view to using it themselves. The task of designing rockets 'as well for war as for triumph' fell to Colonel (later Sir William) Congreve. He was a man of many parts. He had passed through the Royal Academy at Woolwich and worked in the Royal Laboratory there, of which his father was the comptroller; had read for the law; had edited a newspaper; had made many improvements in the manufacture of gunpowder; had built a perpetual motion machine; had developed a new process for watermarking paper; had designed a steam engine; and had been the first to suggest the use of armour-plating on ships.

He carried out his research at Woolwich and by 1805 he had built his first rocket, describing it as ammunition without ordnance and the soul of artillery without the body. The rocket had a cast-iron case, 1 m long and 9 cm in diameter, fitted with a guide-stick 5 m long; it had a range of 1,800 m and it carried an incendiary missile. The rockets performed well when demonstrated on Woolwich Marshes before the Prince Regent and the Prime Minister, William Pitt, and they were used in 1806 to attack shipping collected by Napoleon in Boulogne harbour preparatory to an invasion of England. Some 200 rockets were fired from eighteen ships fitted with special launching gear. The effect was quite unexpected. None of the ships was hit because the wind blew the rockets off course but they landed on the town and set much of it on fire. On a larger scale, Copenhagen was razed by an attack from 25,000 rockets in 1807, and similar attacks were made on Danzig and Walcheren. Rockets were also used with some success in the field at the battle of Leipzig in 1813. But they are probably best remembered nowadays for their use against Fort McHenry, near Baltimore, in 1814, for it was there that Francis Scott Key wrote, in the 'Star Spangled Banner' – now the national anthem of the United States – of 'the rockets' red glare, the bombs bursting in air'.

By then, Congreve had succeeded his father as Comptroller of the Royal Laboratory and developed a range of different rockets, labelled according to their weight, and firing incendiary or explosive missiles. The 32-pounder, which was the most common, had a range of 2,750 m. The British Rocket Corps, with all the associated equipment, training and organization, was established in 1814 by command of his Royal Highness the Prince Regent, and many other European countries followed suit. The use of rockets had already given some advantage to Britain in various campaigns and a bright future seemed to lie ahead. But it was not to be, and the launch of the new Corps was, in the end, something of a damp squib.

Congreve's rockets were light to handle and could be fired very easily, without any recoil, either singly or en masse, and either from the ground or from ships. They had a longer range than the artillery of the day, and they matched it in accuracy, though that wasn't saying much for the rocket with its long stick trailing behind was severely affected by winds and never had a very steady flight. At the battle of Waterloo in 1815 Captain (later General) Cavalié Mercer,

The use of rockets by infantry against cavalry (top); in covering the storming of a fortress (centre); and from ships (below). From Details of the Rocket System *by Colonel Sir William Congreve, 1814. (The Royal Artillery Institution)*

saw the guns standing mute and unmanned, whilst our rocketeers kept shooting off rockets, none of which ever followed the course of the first; most of them, on arriving about the middle of the ascent, took a vertical direction, whilst some actually turned back on ourselves – and one of these, following me like a squib until its shell exploded, actually put me in more danger than all the fire of the enemy throughout the day.[8]

That was probably something of an exaggeration, but the uncertainty involved with rocketry didn't greatly endear it to the Duke of Wellington, who regarded it as a lot of nonsense. Before the battle of Waterloo he had ordered that all the rockets of the second troop, commanded by Captain (later General) Whinyates, should be put into store and replaced by guns. On being told that this would break Whinyates's heart, he replied 'Damn his heart, sir; let my order be obeyed.'[9] Some compromise must have been struck for Whinyates led his troop into action with rockets and six guns. That the rockets did not make any great impact was not due to any lack of bravery on his part for he had three horses shot from under him, and was severely wounded in the arm and the leg, all in one day.

It was the beginning of the end for those early rockets. Wellington's dislike of them cannot have helped their cause, but it was the technical improvements in the gun that eventually gave it supremacy over the rocket. First Boxer and then Hale improved on Congreve's rocket design, the latter by cleverly arranging that the flow of gas from the rear of the rocket caused it to rotate in flight so that it travelled on a steadier path without the need of a stabilizing stick. But during the second half of the nineteenth century, the rocket disappeared as a weapon of war. It was, however, only the end of a chapter; certainly not the end of the rocket story.

* * *

The Congreve rocket was not a great military success, but there was much spin-off in peaceful uses, particularly in improved ship rescue equipment. Until the start of the eighteenth century many lives were lost when sailing ships ran aground on rocky coasts and the distress of those on board could be witnessed from the shore by helpless onlookers. As early as 1791 Lieutenant John Bell had devised an apparatus for throwing a lifeline from ship to shore using a mortar fired by gunpowder, but it was equipment designed in 1808 by George Manby and by Henry Trengrouse that eventually came to be used. Both these inventors had actually seen shipwrecks. Manby had been present when the *Snipe* gun brig sank off the Yarmouth coast, sixty-seven people drowning within 60 m of the shore; Trengrouse had witnessed the wreck of the *Anson* frigate in Mount's Bay in Cornwall with the loss of over a hundred lives.

Manby adopted Bell's idea of firing a lifeline from a mortar but fired it from shore to ship instead of the other way round. Once the line was caught on the ship a hawser was drawn aboard and a cradle could then be run along it to land survivors. Manby's mortars were installed at a number of shore stations and he was paid £2,000 for his invention. The system saved many lives but it was of limited value for it was of no use if there was no shore station close to the site of the shipwreck. Trengrouse's system, which reverted back to firing the lifeline from ship to shore, overcame this problem. It used a rocket to fire the lifeline and was portable enough to be carried on any size of ship. Yet for some reason, the idea did not catch on at all quickly and Trengrouse, who had devoted much of his life and resources to perfecting his project, was paid only £50 by the government and 30 guineas by the Society of Arts even though Trinity House had recommended, in 1825, that every ship should carry one of his rocket systems.

An improved rocket, based on Congreve's design, was patented by John Dennet in 1826, and a still better one was made by Colonel Boxer in 1855. Such lifeline systems have saved thousands of lives all over the world. Other successful ventures were the use of rockets in whaling harpoons, and in firing light, sound and smoke signals.

More fanciful attempts were made to construct rocket-propelled torpedoes and flying machines. Claude Ruggieri, a famous Italian manufacturer of fireworks, was probably the first to use rockets for space travel in the early nineteenth century. He fired mice and rats into the air by rockets, arranging for their return by parachute. It is reported that he even envisaged sending up a small boy until the police intervened. In Britain Charles Golightly – a good name for a rocketeer – invented a steam-driven rocket known as a Steam Horse on which he claimed it would be possible to ride from Paris to St Petersburg in one hour.

But it was not until the limitations of gunpowder as a rocket fuel were realized that the modern era of rocketry and the science of pyrotechnics really began. A young Russian teacher of mathematics, Konstantin Tsiolkovsky, had written an article on space travel in 1895, emphasizing the importance of the speed with which the exhaust gases left the rocket, and pointing out that it would be necessary to replace gunpowder by liquid propellants to achieve the greater thrusts necessary to propel rockets over greater distances. But that was more easily said than done in 1895 and it was not until 1926 that Robert Goddard launched what is generally regarded as the first space rocket, in America. It used a fuel mixture of gasoline and liquid oxygen and though it only flew for a few seconds, it marked an important stage in the history of space travel. Further experiments, directed by the Romanian-born physicist Hermann Oberth in Germany and by Wernher von Braun first in Germany and then in the United States, produced today's enormous space rockets, capable of taking men to the moon and back, most commonly using kerosene or liquid hydrogen, together with liquid oxygen, as fuel.

*   *   *

Guns, like rockets, grew out of fireworks. Gunpowder was being used in China around the year 1000 for propelling various missiles from tubes made first from hollowed-out bamboo stem and later from treated paper (another Chinese invention). The guns were of little use as weapons because the tubes were neither very cylindrical nor very strong, and the missiles did not fit closely within them.

Guns as we know them today, with cylindrical metal barrels and close-fitting

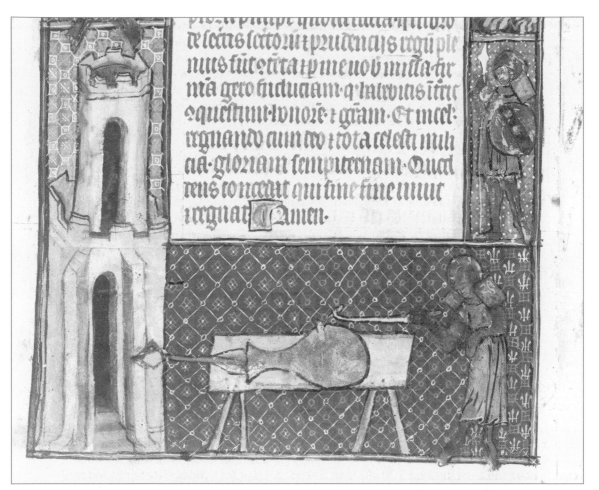

One of the earliest guns. A drawing from the manuscript De Notabilitatibus, Sapientis, et Prudentia *(1326) by Walter de Milemete. (The Governing Body of Christ Church, Oxford)*

projectiles, appeared in China around 1280 and in Europe some fifty years later. There is an illustration of one in *De Officiis Regum*, written by Walter de Milemete about 1325 and owned by Christ Church College, Oxford. It is shaped like a bottle with a long neck. In use, gunpowder was rammed into the wider part, which had a vent leading out of it; the missiles, bolts and arrows such as had been fired successfully from crossbows, were wrapped in leather to ensure a good fit and placed in the neck of the bottle; and the guns were fired by touching the train of powder passing through the vent with a red-hot rod.

Similar guns began to be used in warfare during the fourteenth century; they became effective in the fifteenth and were well established during the sixteenth centuries. They had a revolutionary impact. As Robert Boyle wrote in 1664: 'The invention of gunpowder hath quite altered the condition of Martial Affairs over the world, both by sea and land.' And there were consequential effects on all aspects of contemporary living.

The guns – or bombards as they were commonly called – were first used in land warfare in Europe in 1320. A chronicler[10] refers to

> . . . crakys wer of war,
> That thai befor herd nevir eir

being used by the English in Scotland in 1327. It seems very likely, too, that the English had at least two guns at the battle of Crecy in 1346, at the start of the Hundred Years War, and guns were certainly fired in the same year at the siege of Calais, with round stone shot being used as the missile. A contemporary ballad began with the words:

> Gonners to schew their art
> Into the town in many a parte
> Schot many a fulle great stone.[11]

Such primitive guns cannot have been very effective weapons, though the noise they made might well have had a significantly frightening effect on any enemy. They could not, however, propel a missile of any weight for any great distance, and Calais held out for eleven months until famine forced a surrender. More effective use of guns only became possible about a hundred years later when the art of casting huge barrels in bronze or iron was perfected. The round shot may have had little effect on Calais in 1346, but throughout 1449 the King of France, with the finest artillery in the world at that time, was battering down the remaining English castles in Normandy at the rate of five a month, quickly bringing the Hundred Years War to an end. Twelve years later, during the Wars of the Roses in England, the Earl of Warwick, using 56 cm calibre guns which had been carried to Newcastle by sea, destroyed Bamburgh Castle within a week. And in 1523 Landstuhl, regarded as the strongest fortress in Germany, was flattened in a day by forces under Philip of Hesse.

The vanquished suffered most but the conquerors had a price to pay because their guns could not be relied upon to damage only the enemy. Nor did they show any respect for rank. Taking advantage of the civil strife in England, James II of Scotland laid siege to Roxburgh Castle in 1460, only to be killed, at the age of thirty, by one of his own cannon. A contemporary account[12] describes his death: 'While this prince, more curious nor became the Majestie of any Kinge, did stand near-hand where the Artylliare was discharged, his thigh-bone was dung in two by a piece of a miss-framed gun that brake in the shuting, by the which he was stricken to the ground and died hastily.' Many others met similar, but less well–recorded fates, and much work had to be done before gunpowder could be brought under better control.

* * *

The military use of the power of gunpowder, however imperfect, had important social consequences. Feudalism was the general way of life in the West with the master in his castle controlling everything and everyone around him. The system was well entrenched and seemed likely to last for ever because the castles were well-nigh impregnable to the attacks of ancient weapons such as the perrier, battering ram, trebuchet and ballista. But that was changed by the gun, for fortresses which could previously only be starved out by long sieges could now be smashed to pieces within a few days. Churchill expressed it very trenchantly in his *A History of the English-Speaking Peoples*:

> Amid jarring booms and billowing smoke, which frequently caused more alarm to friends than foes, but none the less arrested attention, a system which had ruled and also guided Christendom for five hundred years, and had in its day been the instrument of immense advance in human government and stature, fell into ruins. These were painfully carted away to make room for new building.[13]

Thomas Carlyle wrote that 'gunpowder made all men tall'.

Many aspects of everyday life changed. One of the major differences was that warfare became much more professional. The steel-clad knight and the bowman began to fade away, and it was no longer effective for local barons simply to arm their serfs with pikestaffs as and when necessary. Guns were expensive to make and difficult to handle, so that a separate military profession and, eventually, standing armies arose.

* * *

Around the world other empires were collapsing, and conquest and colonization were becoming the order of the day. The Byzantine Empire, founded by Emperor Constantine I, Constantine the Great, with Constantinople (now Istanbul) as its capital, survived onslaughts from Turks, Magyars and Saracens for over a thousand years, because it could defend itself. That defence depended largely on the remarkable triple-walled defences of Constantinople, which repelled any land-based attacks; on the strength of the Byzantine fleet which guarded the sea-wall approaches; on the discipline, training and morale of the archers; but perhaps above all on the mysterious but deadly Greek fire.

But in 1453 Greek fire was on its way out and gunpowder was on the way in. Mohammed II, the Sultan of Turkey, one of the first great artillery men, employed a Hungarian engineer called Urban to build two mammoth bronze cannons. They took three months to cast, were over 8 m long, had a bore of about 0.9 m, and could shoot a 320 kg shot a distance of 2.5 km.*

---

* The Dardanelles gun, on show at the Fort Nelson Museum in Hampshire, is 5.2 m long, has a bore of 635 mm, weighs 17,273 kg, and fired a stone shot weighing almost 272 kg. It was built for Mohammed II by Munir Ali in 1426, and it was originally part of the battery of guns defending the Dardanelles. While still in position there in the 1860s, a granite ball was shot from the gun using a charge of 150 kg of gunpowder. The ball broke up after flying for 550 m, scattering fragments across the Dardanelles for a distance of over a mile. The gun was presented to Queen Victoria by the Sultan Abdul Azis in 1867.

The Dardanelles gun. This giant bronze gun was made in Turkey in 1426. (The Board of Trustees of the Royal Armouries)

All was ready for the attack on Constantinople on 6 April and, although one of the two huge guns blew itself up, the walls of the city were seriously breached on that first day, though the defenders were able to repair them during the night. The battering began again on 11 April and continued for six weeks. The huge gun constantly slipped on its foundations and the cycle of filling the chamber with pounds of gunpowder, ramming it home, loading the shot through the muzzle, firing and cleaning was so lengthy that it was only possible to fire seven shots in a day. In the end, that was enough and Constantinople fell on Tuesday 29 May 1453. The Byzantine Empire ended its 1,100-year life, and the Ottoman Empire, which was to survive until 1924, was born. The twin forces of modern warfare – incendiaries in the form of Greek fire and explosives in the form of gunpowder – had met, and gunpowder had won.

Other Turkish guns enabled Baber, the ruler of Kabul, to capture Kandahar and to advance through the Khyber Pass with an army of 12,000 men on to the northern plains of India in 1526. The forces opposing him had about 100,000 men supported by 1,000 elephants but they were no match for his gunpowder and he won the battle of Panipat on 21 April to begin the dynasty of the great Moghuls which ruled northern India for three centuries.

In the same year, far to the west, Suleiman the Magnificent was using other Turkish guns to end the independence of Hungary at the battle of Mohács. Seven years later, the Incas in Peru were overthrown by a small force under Francisco Pizarro, and this, together with the defeat of the Aztecs in Mexico by Hernando Cortez in 1521, meant that the whole of South America and some of the Southern States of North America fell to but a handful of Spaniards. In their turn, other native peoples, such as the Aborigines in Australia, the Maoris in New Zealand, many tribes in Africa, and the Red Indians of North America, suffered the same fate.

Would any of it have been possible without gunpowder?

*   *   *

Guns were first used at sea in the battle of Sluys in 1340, which was fought to prepare the way for the English army's advance into France. Edward III, known in his early days as the 'King of the Sea', had begun to equip the Navy prior to the Hundred Years War. Export of timber suitable for ship-building had been prohibited and by 1338 at least three ships had been fitted with a few guns. They did not contribute significantly to the outcome of the engagement at Sluys – that was won by the British archers – but what a portent.

By 1373 there were frequent references to powder and shot in naval documents and, as ships began to grow in size, guns came to play a more and more important role. At first, land guns were mounted on to wooden structures known as 'castles', which were built on the deck of a vessel in the bow or the stern, but later guns specially designed for use at sea were placed on gun decks and fired through holes in the ship's side. By the end of the sixteenth century a wide variety of naval guns had been fitted into many different types of ship. They ranged from 20 cm bore cannons royal – 2.5 m long, weighing 3,600 kg and firing 30 kg shot – to 25 mm bore robinets – weighing 135 kg and firing 0.2 kg shot. In between, there were cannons, cannons serpentine, bastard cannons, demi-cannons, cannons pedro, culverins, basilicos, demi-culverins, bastard culverins, sakers, minions, falcons, falconets and serpentines.

Both guns and ships were severely put to the test when the Spanish Armada – 125 ships carrying 300,000 men – appeared off the Cornish coast on 31 July 1588. Sir John Hawkins, one of the English admirals, described it as 'the greatest and strongest combination . . . that ever was gathered in Christendom'.[14] In tonnage and manpower it was almost twice as strong as the force which England could muster to oppose it.

Both fleets were very heavily armed. The Spanish carried about 2,500 guns with 125,000 rounds of ammunition (an average of 50 rounds per gun). Figures for the English fleet are much less precise, but there were probably about 2,000 guns with, at first, about 50,000 rounds of ammunition (25 rounds per gun). There was thus no lack of fire-power on either side and, in the four engagements which were fought as the Armada moved relentlessly along the length of the

English Channel, the guns were constantly in action. So much so that Charles, Lord Howard of Effingham, the Lord Admiral, on board his flagship *Ark Royal*, referred to 'a terrible value of great shot'[15] being expended, and both sides feared that they were going to run out of ammunition.

Fortunately, the English gunners did better than the Spanish. They were more experienced and their guns were easier to handle so that they could maintain a consistently higher rate of fire. But it was not only in gunnery that the English had the advantage. They were not fighting 500 miles from home; they had leaders with recent experience of naval battles; their ships were better provisioned, less crowded and much cleaner; their seamanship was second to none; and they adopted better tactics. The Spanish showed no lack of bravery, but they relied on the traditional tactics of closing with the enemy and boarding his ships. Philip II had indeed ordered his commanders to 'bring the enemy to close quarters and grapple with him'.[16] This was why the Spaniards relied on heavy guns (some of them weighing 2.5 tonnes), carried large numbers of soldiers on their vessels, and went into action in a huge crescent formation.

The English, by choosing to fight the battle at long range rather than at close quarters, heralded a new era in naval warfare. Their mobile, nimble ships were used to outflank the enemy, to keep on the weather-side and to pick off any stragglers. The ships' guns were used to inflict constant damage to wear the enemy down. 'We pluck their feathers little and little'[17] was how Lord Howard put it.

The plucking took rather a long time and it was only the dramatic last-minute intervention of fire-ships that tipped the scales. It was probably the first but certainly not the last time that incendiaries were to prove themselves a more potent weapon than explosives. The use of fire against ships was not new. Greek fire had blazed a long trail, and in 1585 a Dutch attempt to relieve Antwerp from a Spanish blockade had been assisted by an Italian engineer called Frederico Giambelli, who designed some ingenious floating bombs and fire-ships. These were sent down the River Scheldt against the Spanish positions, which were centred on a huge 0.8 km long wooden bridge built across the river. The Spanish held out, but Giambelli's 'infernal machines' caused much damage and confusion and killed almost a thousand people. They passed into history as the 'hellburners of Antwerp'.

By chance, Giambelli had moved to London to supervise the building of a boom across the Thames at Gravesend and his services were recruited when the Navy decided to use fire-ships. Eight old vessels, of around 150 tonnes each, were gathered together, stripped, stuffed with anything and everything that would burn, and had their guns loaded with double shot. Tide and wind were favourable for an attack on Sunday 7 August, and at midnight the floating flaming inferno fell upon the Armada as it lay at anchor off Calais. The blazing hulks, coupled with the explosions of the guns as they got hot, caused such panic that the disciplined order of the Spanish fleet was at last broken. At dawn on the following day their commander tried to re-form his demoralized fleet, but, after a nine-hour running battle along the coast off Gravelines, defeat stared him in the face and his scattered ships retreated into the North Sea and set sail for the hazardous journey back home around the north of Scotland. England once again ruled the waves, but

Spanish and English ships at the battle of Gravelines, after the Dutch artist Visscher. In this version Calais and Dover have been transposed making it appear, incorrectly, that the Armada is fighting its way back through the English Channel. (National Maritime Museum)

it had been a much closer thing than the popular legend of Drake and his game of bowls suggests.

The post mortem raised many issues but on one point there was general agreement – the impact of the guns, on both sides, had been disappointing. The star-billing of their exotic names had belied their real performance. A contemporary commentator summed it up when he wrote 'so much powder and shot spent, and so long time in fight . . . and so little harm'.[18]

* * *

There were a number of reasons why early guns did not live up to their expectations. They could only be charged and fired very slowly; if they were heavy they were difficult to handle, particularly on board a ship; and loading through the muzzle meant that guns on ships had to be retracted before reloading.

Gunpowder varied greatly in quality and was very susceptible to damp; it was not cheap, so gunnery training was a costly business; and the powder produced great quantities of nasty, dense smoke when it exploded, which made firing guns in the confined space of a gun-deck a particularly dirty business. Still more important, the guns were so unreliable and the solid iron or stone shot which they propelled could only inflict very limited damage on a target of any strength.

The gamble of firing almost any gun during the Armada, and for some years after, was well described by Professor Mattingly when he wrote:

> not only were no two cannons ever quite alike but the cannon balls supplied with any given piece were unlikely to be all the same size, so that the 'windage',[19] the difference between the diameter of the shot and that of the bore, usually considerable, was also variable. As a result, it was only in the text books that a piece of given bore and length loaded in a given fashion would hurl a ball of a given size a given distance. In fact, even the most experienced gunner might hesitate to predict whether when he next fired it his gun would send its shot directly to the target, drop it with a discouraged burp a few hundred feet ahead, or blow up at the breach, probably killing him and his crew. At long ranges the chances of effective execution were slight.

So it was with some trepidation that, in Shakespeare's words,[20] 'the nimble gunner with linstock* now the devilish cannon touches'.

Even when the cannon did perform well it was singularly disappointing to find its missile doing little harm, and it had long been the dream of artillerymen that they might be able to fire a projectile which would itself explode on contact with its target. That dream began to come true around 1500, when stronger gun barrels enabled solid cast-iron round shot to be fired instead of the lighter stone shot. The metal shot could then be made into a variety of patterns – chain shot, case shot or grape-shot – all designed to increase its destructive effect. More importantly, it could also be made in the form of a hollow sphere, which could be filled with an incendiary mixture and fitted with holes to take a fuse and to allow the flames to escape. The incendiary shell had been born. One of the earliest types was invented by Valturio in 1460, and so-called carcasses were widely used by the British between 1700 and 1870. They were filled with a mixture of saltpetre, resin, sulphur, turpentine and tallow, and fitted with a fuse, like a quick-match, made by boiling cotton wick with a mixture of potassium nitrate, charcoal and gum, followed by careful drying. The fuse was lit by the flash from the propellant as the gun was fired and the time of burning depended on the length of fuse used.

It was only a short step from here to filling the carcass with an explosive – some of the early carcasses had, indeed, blown up instead of burning – and the only explosive available was gunpowder. But that small step could not be taken

---

* A linstock held a burning slow-match for igniting the gunpowder in a gun. For artillerymen on land the match was commonly held in a sideways extension at the end of a pike.

quickly because it was difficult to devise a satisfactory fuse. It did not matter much when an incendiary shell ignited because it went on burning for some time, but it was important that an explosive shell did not blow up until it hit its target. If it exploded in the gun barrel it damaged the gun; if it blew up in mid-air it was simply wasted.

Getting the timing right, especially at first, was not easy and one of the gunner's favourite methods of doing it was to recite the Apostles' Creed; it at least enabled him to hedge his bets. It was not until the invention of the watch in 1674 that more accurate measurement of time became possible, and thereafter the use of explosive shells became much safer and more popular. But old and new were still used alongside each other for many years to come. At the siege of Gibraltar by the Spaniards between 1779 and 1783, the British defenders fired nearly 130,000 explosive shells, but they also used almost 1,000 carcasses and 58,000 round shot, including some that were heated before firing to try to inflict greater damage on the enemy ships. They needed 8,000 barrels of gunpowder.

Most of the other disadvantages of the early guns were overcome during the nineteenth century. Slow-burning fuses were replaced first by clockwork devices and by 1850 by percussion or concussion fuses. Percussion fuses operated by the shock of impact with the target, concussion fuses by the shock of discharge on firing. By this time both the gun and the shell had been extensively redesigned. Muzzle-loading had been replaced by breech-loading which enabled a much more rapid rate of fire; the barrels had been strengthened at the breech end to withstand greater pressure; mechanical arrangements for taking up the recoil of the gun had been designed; better gun-carriages were available which gave improved mobility; and gun barrels had been rifled in order to impart a spin to the projectile as it was fired, to ensure greater accuracy in flight. To enhance this, the spherical-shaped shell had been replaced by the modern, elongated streamlined shape. The commonest fuse was a percussion fuse in the nose of the shell.

Gunpowder, which had served both as the propellant and shell-filling, but which had never been by any means ideal for either purpose, was superseded in the former role by smokeless powders, such as cordite, in 1890, and in the latter by new nitro-compounds in 1896. Modern artillery, which made the old guns look distinctly archaic, had arrived and its achievements in the twentieth century made all previous events pale into insignificance. Gunpowder had been powerful in its day, but it was something of a damp squib when compared with its replacements.

* * *

Although it first appeared in the West at the same time as the cannon, the hand-gun had to serve a much longer apprenticeship before it made any appreciable impact, for almost three centuries passed before it began to challenge the supremacy of the bow and arrow. This old weapon is not very highly regarded nowadays but in its day it was very formidable. By the fourteenth century a trained archer using a 1.8 m bow could fire rapidly, accurately, quietly and cleanly and the arrows could penetrate armour at a range of well over 230 m. At the battle

Old handguns, from a Chinese text by Wu Pei Chih, 1628.

of Crecy, Froissart writes that arrows were shot 'with such force and quickness that it seemed as if it snowed'[21] and, with only a few thousand archers, the English came close to conquering the whole of France within a hundred years.

So the first hand-gun had a lot to contend with. It consisted of a short, cannon-like barrel fastened on to the end of a wooden stick, which served as a simple stock. The barrels were smooth-bored and loaded through the muzzle, and the guns were fired in much the same way as a cannon by applying a red-hot wire, or, later, a slow-match, to the touch hole at the end of a vent in the lower part of the barrel. They had a range of only about 45 m; they were very inaccurate; they had a very slow rate of fire; and they produced a lot of smoke and dust. To aim and fire them at the same time was well-nigh impossible for one man, and they were so unreliable that the gunner was rather more at risk than his enemy. The only way in which they were superior to the bow and arrow was that they made a lot of noise, and this was particularly frightening to horses.

Yet their potential was obvious and it was only a matter of modification and time before they became effective weapons. The rudimentary stock was reshaped so that it would fit under the arm or against the shoulder. Serpentine gunpowder was replaced by corned powder and special blends particularly suitable for small guns were mixed. Correct portions of powder and, later, a bullet were sealed in a cartridge made from stout paper sewn up at both ends and waxed to keep out moisture. Hence the stationer's term 'cartridge paper'. A bayonet was fitted to the end of the barrel so that the ancient pike could be done away with. But most importantly, the gun was made easier to fire by fitting a firing- or flash-pan, which could hold some fine gunpowder, as priming, on the outside of the barrel adjacent to the touch-hole, and a number of mechanical ways of igniting this priming were invented. These enabled the firer to hold the gun with both hands instead of having to leave one free for holding the slow match.

The matchlock system, dating from around 1410, was the first. A slow match was held in an arm so that its glowing end was positioned just above the priming in

A seventeenth-century musketeer.

the pan. On pulling the trigger of the gun, the arm was released so that the priming was ignited and, it was hoped, the gun fired. The term 'a flash in the pan' referred to a misfire in which the priming flashed without setting the gun off. In the wheel lock system, invented in Germany around 1520, pulling the trigger released a spring-loaded wheel which had previously been wound up. It was so arranged that its serrated edges rubbed against a piece of pyrites and the resulting sparks ignited the priming. By 1610 the flintlock had emerged; this produced sparks by causing the sharp edge of a flint to strike against a roughened plate.

The matchlock system was the cheapest and simplest and was the first to be adopted by European infantry, but it did not play a particularly significant role in warfare until the time of the Civil War in England, around 1650. The very early guns had been improved by then but they still suffered from the severe disadvantage that it was extremely difficult to keep a slow match constantly alight during all sorts of weather, and Cromwell's reported advice to his troops – 'Put your trust in God, my boys, and keep your powder dry'[22] – was, indeed, very apt.

The use of hand-guns for sporting purposes had become very popular during the sixteenth century and to achieve the greatest accuracy and ease of firing, those sportsmen who could afford to spend large sums on their individual weapons soon turned from the matchlock to the wheel lock and the flintlock, and also to the rifled barrel, first introduced in 1520. And in time of war, sporting guns were commonly called upon to arm snipers and sharpshooters. This enabled the military to see at first hand what the superior guns could do and around 1700 matchlock guns were superseded by flintlocks in most European armies, though they continued in use in other parts of the world for many more years. In England the long life of the Brown Bess musket began.

Brown Bess (the origin of the name is something of a mystery) was a flintlock gun with a 1.2 m barrel and a 19 mm bore. It had a 0.43 m long bayonet and shot rather loose fitting bullets weighing 28 or 30 to the kilogram. It became a great favourite with most soldiers and was used by the Duke of Marlborough at the battles of Blenheim and Ramillies, in the conquest of much of the British Empire, and at Waterloo. Yet it was by no means ideal. Loading and firing it was certainly something of a business, requiring the sort of careful sequence of actions that remain the basis of all modern arms drill. The musketeer first bit the end off the cartridge, squeezed a little gunpowder into the firing-pan, and poured the rest down the barrel. He then inserted the bullet into the barrel, pushed the remaining paper on top of it, and rammed the whole lot down with a rod. He could fire about two or three rounds per minute, but he could not keep that rate up for very long, as the flint had to be changed every thirty shots and the barrel needed cleaning at least every hundred shots.

Firing the musket was, indeed, a dirty job. Much of the powder remained as a deposit in the barrel and on the firing-pan, and dense clouds of smoke were produced, so much so that the whole battlefield was shrouded in smoke with visibility frequently less than 45 m. That meant that the musketeer could not always see what he was supposed to be firing at and he had to hold his fire accordingly. This was also necessary because the gun was so inaccurate. The bullet could kill a man at a distance of 450 m but hitting anyone at that range was purely accidental. The effective range was only about 75 m and aiming at a man 180 m away was like baying at the moon.

But, with all its shortcomings, Brown Bess was distinctly superior to the French weapons and the tactics of the British generals were designed to get the best out of it. Above all, it was very popular with the troops and, although there were some diehards who called for the return of the bow and arrow after Waterloo, the majority felt some sadness when the Brown Bess was withdrawn from service shortly before the Crimean War. It had been the queen of the battlefield for a century and a half; gunpowder her faithful servant.

The smooth-bored barrel disappeared and the new generation of guns were rifles. Sportsmen had long since demonstrated that a gun with a rifled barrel which imparted a spin to the bullet was much more accurate than a smooth-bored musket, and Benjamin Robins, who designed the ballistic pendulum, had written an impressive paper in 1747, entitled 'Observations of the Nature and Advantage of Rifled Barrel Pieces'. It concluded that 'whatever State shall thoroughly comprehend the nature and advantage of rifled barrel pieces and . . . shall introduce into their armies their general use . . . will by this means acquire a superiority, which will almost equal anything that has been done at any time by any one kind of arms'.[23] The stakes were high.

By that time, American troops had already been issued with a rifle, designed by immigrant German gunsmiths, and they used it very effectively against the British in the War of Independence. This led to the establishment of the British 'Experimental Corps of Riflemen', later to become the Rifle Brigade, who were equipped with the Baker rifle at Waterloo. It was so much more accurate than the Brown Bess that it was clear that rifles were here to stay, even though they had, at first, an even slower rate of fire than Brown Bess. This arose because it was necessary to have a tight-fitting bullet to get the maximum effect from the rifling in the barrel. To achieve this in the Baker rifle, the ball had to be placed in the centre of a greased piece of leather or rag before it was rammed in on top of the powder. At first, wooden mallets were issued to facilitate this operation, but they were soon discarded as impractical.

The problem was overcome by an extensive redesign of the bullet. It was elongated so that it was pointed at one end and had a hollow base containing an iron cup at the other. When it was fired in the barrel, the iron cup was driven up into the base of the bullet causing it to expand so that it fitted the barrel tightly. The main credit for this invention is given to Colonel Minié, a Frenchman, and the British government paid £20,000 for the rights to use the idea. Rifles using the Minié-type bullet were adopted as general issue by the British Army in 1851,

Some nineteenth-century rifles. Top to bottom: *the Minié rifle (1851); the Enfield rifle (1853); the Artillery rifle (1853); Lancaster's oval-bore rifle. (The Board of Trustees of the Royal Armouries)*

The SA80: the standard rifle of the British Army, seen here with an image-intensifying nightsight. It weighs 4.98 kg and is 750 mm long. It has an effective range of 500 m and can fire at a rate of 510–770 rounds per minute. (Crown Copyright: the Army Picture Library)

and these in turn were replaced soon after, in 1853, by the Enfield rifle. This was widely used in the Crimean War and by both sides in the American Civil War (1861–5). It acquitted itself well, and it is unfortunate for its reputation that it is best remembered nowadays as the immediate cause of the outbreak of the Indian Mutiny. A rumour went around that the cartridges issued with the gun had been waterproofed using grease made from cow's fat and the lard from pigs. As cows are sacred to the Hindus and pigs anathema to Muslims, and as the rumour could not be quelled, the results were disastrous.

By this time the flintlock system of firing had been largely replaced by percussion firing. An ardent wild-fowler, the Revd Alexander Forsyth, concerned that the sparks from a flintlock alarmed his targets, had discovered that it was possible to achieve a flash which would ignite the gunpowder, by striking a particular mixture of chemicals a sharp blow, and this soon proved to be a much more reliable method of firing a gun than any previous one. It had been adopted by many sportsmen by 1820 and came into more general use around 1845.

Other changes introduced in the latter half of the nineteenth century – breech-loading, repeat firing and automatic loading from a magazine – brought the rifle to something approaching its present design, as represented by the Russian Kalashnikov, the American Armalite, the French FA MAS, the British SA-80, and the Belgian FN FAl, adopted by NATO. These go on firing as long as the trigger is held down so that they produce a heavy volume of shot, but they have a shorter range and are less accurate than many of their predecessors. There were similar advances in the design of pistols, revolvers, automatics, machine-guns and sub-machine-guns, commemorated by such names as Colt, Borchardt, Lüger, Derringer, Browning, Webley, Smith & Wesson, Gatling, Maxim, Lewis, Vickers, Hotchkiss, Bren, Sten, Thompson ('Tommy') and Sterling.

All this activity meant that by the turn of the century every leading military power had a wide range of small arms available. Along with the improved artillery, they brought the supremacy of cavalry to an end and they turned large-scale frontal assaults by infantrymen into suicide missions. These lessons were learnt, at enormous cost, during the First World War when it slowly began to be realized that it was necessary to build not only trenches in the ground but also vast fortifications and armoured vehicles. The nature of warfare had changed. And the vast array of new guns was beneficial to the sportsman. Alas, too, to the criminal.

CHAPTER 6

Mining and Civil Engineering

The impact of gunpowder on warfare was paralleled, if not outdone, by its effect on the peaceful activities of miners, quarrymen, tunnellers and civil engineers. Considering the enormous contributions that they have made to the advancement of civilization their use of gunpowder, together with that of automatic rock drills, must rank among the greatest innovations of all time.

It is not easy to split hard rock, which is necessary in the extraction of most metallic ores, and the methods available before the use of gunpowder were severely limited. In fire-setting, the rocks were heated by lighting a bonfire against them and then drenching them with cold water. This method had been adopted by Hannibal when he was passing over the Alps with his elephants, though he used vinegar rather than water. It is, at best, singularly haphazard, and it is very slow. When carried out underground it is also very dirty, filling the working space with clouds of smoke and fumes. It was, however, such a cheap method that it was still practised in some backward mines until well into the eighteenth century.

The more reliable, but more difficult and more expensive, method involved boring holes into the rock and then splitting it open by wedges. Typically, a two-man team, one holding a steel borer with a specially shaped and treated head and the other wielding a hammer, might take three hours or more to drill a hole 50 cm deep, with constant stopping and starting because the borer only lasted about fifteen minutes before it had to be returned to the smithy for re-tempering. When the hole was deep enough, two semi-cylindrical rods of iron or steel, known as feathers, were positioned in the hole and a steel wedge was driven in between the flat sides of the two feathers. Hopefully, the rock would split. If not, it was a case of try, try and try again. It was very hard labour.

In lime-breaking, the hole drilled into the rock was partly filled with fresh quicklime, some water was added, and the hole was closed with a tight-fitting wood plug. The heat generated by chemical reaction between the lime and the water produced some steam and the hope was that the build-up of pressure would be sufficient to split the rock. It was a very slow, clumsy and precarious method.

But gunpowder was to change all that. As the poet Dr Dalton wrote in 1753,

> Dissever'd by the nitrous blast,
> The stubborn barrier yields at last.[1]

Modern quarrying with air-decking (p. 90). Some 35,000 tonnes of well-fragmented limestone is being blasted away from the 17 m wide face of a Gloucestershire quarry. Twenty-six drill holes with a diameter of 140 mm were charged with packaged emulsion and ANFO, with air-decking. (Photograph by Julian Cleeton)

The earliest record of the use of gunpowder for blasting in a mine occurs in the *Proceedings of the Schemnitz Mine Tribunal* for 8 February 1627. This gives an account of a blasting demonstration by Caspar Weindl in a mine in Hungary, and the method soon spread to Germany. It was introduced into Britain when Prince Rupert brought some central European miners to work in the Ecton copper mines on the boundary of Derbyshire and Staffordshire in 1638, and it was being used in Somerset in 1670. An exponent of the art, one Thomas Epsley, was encouraged by the Godolphin family to move from Somerset to Cornwall to train the miners at Breage in the new technique. He arrived in June 1689, but he had been killed, possibly by a gunpowder explosion in the mine (or 'bal' in Cornish) by the end of that year. The Breage parish register records that 'the man that brought that rare invention of shooting the rocks . . . died at the bal and was buried the 16th day of December in the year of our Lord Christ 1689'.[2]

The use of gunpowder in blasting progressed in three stages. At first, until 1831, it was both laborious and hazardous. Laborious, because the shot-hole for holding the gunpowder was still drilled in the old-fashioned way; hazardous, because there was no reliable method of igniting the powder. Typically, a shot-hole, 0.6 m or more long, was bored into the rock and cleaned out by a swab-stick, a wooden staff with its end splayed out to form a rudimentary brush. Powder was poured in if the hole sloped downwards, or inserted on a long, spoon-like rod if it were horizontal. If the hole sloped upwards, a rough sort of cartridge was made by wrapping the powder in paper and sealing the edges with tallow from a candle. A long, thin, tapering rod of iron or copper, known as a pricker or a needle, was then pushed in with the lower end reaching to the bottom of the hole and the upper end protruding from the top. The space around the needle and above the powder was packed (tamped or stemmed) by pressing in clay or sand with a ramming- or tamping-rod. The needle was then carefully withdrawn and the narrow channel which remained was filled with loose, fine-grained gunpowder. This thin trail of powder was ignited by a length of touch-paper or the end of a candle so fixed, usually with clay, as to give the shot-firer some time to run as far as he could from the scene. Alternatively, a long, slender, home-made tube of rush or straw or goose-quills packed with powder was used to provide the trail. A great deal was left to chance, and the number of miners killed or maimed by premature explosions or by the investigation of misfires was horrendous.

The risks were, fortunately, greatly reduced when William Bickford, a Devonian working among Cornish miners, invented Safety Fuse in 1831. It consisted of a rope, about 12 mm in diameter, with a trail of gunpowder down the centre. The great advantage of this fuse was that it had a constant burning time, typically 80–100 seconds per metre, so that a shot-firer could decide how much time he needed to take cover and could cut off the necessary length of fuse from the 8 metre coil in which it was supplied. The fuse was then inserted down the shot-hole into the gunpowder charge and, after tamping, the open end protruding from the hole was lit. The fuse was so superior to earlier methods of firing that, despite its higher cost and the innate conservatism of Cornish miners, it was readily adopted, and within a few years the number of blasting accidents had fallen by about 90 per cent in West Cornwall. But the hard labour of drilling the shot-holes remained until the advent of automatic rock drills, powered by compressed air, in 1867, and by then the days of gunpowder for blasting were numbered.

It had always been a very good blasting agent because of the relatively slow build-up of pressure when it exploded which produced a steady heaving action, but the smoke and fumes formed when it exploded were very unpleasant and harmful, particularly in any underground working. It left an atmosphere of lingering black smoke, deficient in oxygen, with a nasty smell of bad eggs caused by the hydrogen sulphide which was present, and a dangerously high concentration of poisonous carbon monoxide. The working conditions in early, ill-ventilated mines were bad enough without that. The air was already foul from the exhalations of the miners, from the burning of candles and lamps, and from the complete lack of any sanitary facilities. On top of that, mines were generally very wet, and either very cold or very hot. As the job also involved extreme physical exertion, the lot of the miner was miserable, squalid and dangerous. It was also poorly paid and very insecure. It was, in many cases, only the fear of starvation that kept men at work.

*   *   *

The conditions in coal-mines were even worse than those in metal mines because in many of them – the so-called gassy or fiery ones – there was a potentially explosive atmosphere of fire-damp. This was caused by the seepage of small amounts of naturally occurring methane, now well known as the main component of North Sea gas, into the underground workings. In some mines, too, coal dust added to the risk of an explosion taking place.

If a sufficient quantity of fire-damp builds up in a coal-mine it might be set off by a naked light or by a spark, with a resulting disastrous explosion. A contemporary writer vividly described the effect:

> On the approach of a candle, it is in an instant kindled; the expanded fluid drives before it a roaring whirlwind of flaming air, which tears up every thing in its progress, scorching some of the miners to a cinder, and burying others under enormous ruins shaken from the roof; when thundering to the shaft, it converts the mine, as it were, into an enormous piece of artillery, and wastes its fury in a discharge of thick clouds of coal-dust, stones, and timber, together with the limbs and mangled bodies of men and horses. But this first, though apparently the most appalling, is not the most destructive effect of these subterraneous combustions. All the stoppings and trap-doors of the mine being blown down by the violence of the concussion, and the atmospheric current entirely excluded from the workings, such of the miners as may have survived the discharge are doomed to the more painful and lingering death of suffocation from the after-damp or stythe, as it is termed, which immediately results from the combustion, and occupies the vacuum necessarily produced by it.[3]

Quite horrible mine explosions were commonplace at the start of the nineteenth century but that which happened at Brandling Main (or Felling) Colliery, near Sunderland in the north-west of England, on 15 May 1812 was of special significance for several reasons. First, the loss of life was much higher than in any previous incident, with only 29 of the 121 men underground surviving; second, the mine had been regarded as a model of perfection; third, the unhappy events were recorded in dramatic detail by the local vicar, the Revd John Hodgson; and last, it all led to some positive action being taken to prevent similar tragedies.

Hodgson's report was widely circulated and part of it, unknown to him, was reproduced in the *Annals of Philosophy* for May 1813, where by chance it was read by a London barrister, Mr J.J. Wilkinson, 'a gentleman distinguished for the humanity of his disposition'. He was so moved that he proposed the formation of a society for preventing accidents in coal-mines, and a meeting of interested parties took place in Sunderland on 1 October 1813. What came to be known as the Sunderland Society was founded, and coal-miners the world over have cause to be grateful to the work it was able to do.

Its first report was issued in November 1813, and a letter from Mr Buddle, a local colliery manager, to the Chairman referred to 'the hopes of this society ever seeing its most desirable objects accomplished'. It concluded that 'it is to scientific men only that we must look up for assistance in providing a cheap and effectual remedy'.[4] And so it was that the Society decided to get in touch with Sir Humphry Davy at the Royal Institution.

*　*　*

Humphry Davy was of medium height, with light brown, curly hair, a small head, very bright eyes, an aquiline nose, and a lively temperament. He was born in Penzance on 17 December 1778, the son of a wood-carver, and was educated first in Penzance and then, for a final year, in Truro. On leaving school, aged fourteen, Davy did little or nothing for a while, and it was probably the death of his father, who was in debt to the tune of £1,300, that steered him in more positive directions. His mother, with five children to care for, set up as a milliner and took in lodgers, and Humphry was apprenticed, at the age of seventeen, to a local surgeon and apothecary. He grasped this opportunity with great zeal, beginning to read widely, to learn French, and to take much interest in poetry.

He made such a good impression that he was appointed superintendent of the Medical Pneumatic Institute in Bristol when he was only nineteen. Four years later, he was invited to join the Royal Institution, recently founded by Count Rumford, as Director of the Laboratories and Assistant Lecturer in Chemistry. With a salary of 100 guineas a year, plus coal, candles and a room, he foresaw great opportunities, writing, 'The Royal Institution will, I hope, be of some utility to society. . . . Count Rumford professes that it will be kept distinct from party politics; I sincerely wish that such may be the case, though I fear it. As for myself, I shall become attached to it full of hope, with the resolution of employing all my feeble powers towards promoting its true interest.'[5] And so he did. He gave regular lectures which were attended by many distinguished people; he made many discoveries in the field of electro-chemistry; and he acted as an adviser, for a short period, to the Ramhurst Gunpowder Mills in Tonbridge. He was knighted in 1812 and married a wealthy, socialite widow in the same year. He employed Michael Faraday, his eventual successor, as an assistant in 1813; and he became President of the Royal Society in 1826.

*　*　*

Davy was briefed, on behalf of the Sunderland Society, by Mr Buddle. 'I explained to Davy,' he wrote, 'as well as I was able, the nature of our fiery mines, and that the great desideratum was a light that could be safely used in an explosive mixture. I had not the slightest idea myself of ever seeing a solution.' But his pessimism was misplaced, because a smiling Davy replied: 'Do not despair; I think I can do something for you in a very short time.'[6]

He was to be true to his word, for he had, essentially, solved the problem within three months by insulating a lamp from its surroundings. The trick was to cover the light from an oil lamp by a cylinder of fine wire gauze. The metal gauze spread out the heat from the light inside so that the surface of the lamp never reached a temperature anywhere near as high as that of the light itself. Whereas the naked flame was hot enough to explode fire-damp, the surface of the lamp was

A drawing by Michael Ayrton of Sir Humphry Davy holding a safety lamp. (ICI)

always too cool to do so. Any explosive gas mixture passing through the gauze was ignited within the lamp causing the light to burn more brightly or to blow it out altogether. This was an added advantage, for it meant that the lamp served as a detector of dangerous gas mixtures. As a further refinement, a glass cylinder was inserted in the lower part of the lamp, below the gauze cylinder, to provide a higher level of illumination.

The Davy safety lamp, as it came to be called, was tested by Mr Buddle within a fiery mine. He saw that the surface of the lamp became hot, but there was no explosion. 'We have subdued the monster' was his comment, and he wrote to Davy on 1 June 1816: 'After having introduced your safety lamps into general use, in all the collieries under my direction, where inflammable air prevails, and after using them daily in every variety of explosive mixtures, for upwards of three months, I feel the highest possible gratification in stating to you that they have answered to my entire satisfaction.'[7]

Davy refused to patent his invention, writing, 'I have never received so much pleasure from the result of any of my chemical labours; for I trust the cause of humanity will gain something by it.'[8] His reward came in worldwide acclamation and he became a national hero, with honours bestowed upon him on all sides. The coal owners of Tyne and Wear presented him with a service of plate, valued at £2,500; he was awarded the Rumford Medal by the Royal Society in 1818; and, emphasizing the universal aspect of his discovery, he was presented with a superb silver vase, carrying a figure representing the God of Fire weeping over his extinguished torch, by the Emperor Alexander of Russia.

Davy's later years were less happy. He became something of a snob; his relationship with his wife, and hers with Faraday, were singularly unhappy; he was embittered by claims that he had stolen the idea for his safety lamp from George Stephenson, then a wheelwright at Killington colliery, but later the designer of railway locomotives; and he was in poor health. Much of his time was spent travelling around Europe, and early in 1829, when he was in Rome, he was stricken with palsy. His younger brother, Dr John Davy, caring for him with great devotion and skill, decided to move him from the Italian heat to the cooler climate of Switzerland. Alas, the long journey by horse and carriage took its toll and Sir Humphrey died, in Geneva on 29 May 1829. He was only fifty-one. A few months earlier he had written in a letter, 'If I die, I hope that I have done my duty and that my life has not been in vain and useless.'[9]

* * *

Despite Mr Buddle's original enthusiasm and the many tributes paid to Davy, his lamp was not, at first, widely used and the hopes of greater safety that its invention had built up did not materialize. There were in fact more deaths from coal-mine accidents in the twenty years following Davy's invention than in the twenty years before it. The main reason was the introduction of gunpowder into coal-mining alongside that of the safety lamp.

The explosive was first used in the industry around the middle of the eighteenth century but its use at that time was on the surface and in sinking new shafts. It was only later, around 1820, that it began to be used for underground blasting. It was welcomed at first by both miners and mine owners, the latter because it greatly increased productivity and the former because it made their job so much easier. Thomas Wilson contrasted the old with the new in his poem, 'The Pitman's Pay',[10] written in 1826:

> Here agyen had awd langsyners
> Mony a weary, warken' byen,
> Now unknawn to coaly Tyners,
> A' bein' mell-and-wedge wark then.
>
> Aw've bray'd for hours at woody coal,
> Wi' airms myset droppen frae the shouther
> But now they just pop in a hole
> And flap her down at yence wi' pouther.

But it quickly transpired that gunpowder produced a lot of small coal and slack, and, much more important, that it introduced a completely new hazard into coal-mining. If the use of the safety lamp had been made compulsory and gunpowder banned right from the start, things would have been very different. As it was, the better safety standards brought about by the use of the safety lamp were more than cancelled out by the greater danger in the use of gunpowder. It was a classic case of putting profit before safety. Gunpowder improved a mine's productivity, but at a high price.

The matter could only be resolved by legislation, but while that was coming the mine disasters went on and on: 75 miners were killed at Heaton in 1815; 102 at Wallsend in 1835; 52 at South Shields in 1839; and 95 at Hanwell in 1844. Each in its turn was investigated by a long line of committees who produced worthy and wordy reports but there was very little action until 1842 when an Act of Parliament forbade the employment underground of females and of boys under the age of ten. For the first time, too, independent inspectors were appointed to ensure that the provisions of the Act were observed.

Much more legislation followed, ushering in a period during which mining conditions were slowly but steadily improved. In 1846 it was recommended that the use of safety lamps in all fiery mines should be made compulsory. From 1850 all mines had to keep proper plans, all fatal accidents had to be reported, and an increased number of inspectors were empowered to operate underground. The Royal School of Mines was opened in London in 1851, and the first Mining Institute, at Newcastle upon Tyne, in 1852. In 1860 all earlier Acts were amended and made permanent and the general rules of safety were extended still further.

But none of this completely prevented further explosions. Much progress was made in improving the ventilation systems in mines, and there was a better understanding of the dangers caused by coal dust, but gunpowder went on being used and was a constant hazard. The vital step forward came when a Royal Commission on Accidents in Mines, sitting between 1881 and 1887, took the radical step of recommending that, henceforth, only specially blended and tested explosives – so-called safety or permitted explosives – should be used. So the days of gunpowder in coal-mining were over.

* * *

The first time that gunpowder was used on any large scale in a civil engineering project was in the construction of the Languedoc Canal across the south-west corner of France. It was finished in 1681. Linking the Mediterranean Sea with the Bay of Biscay, the canal was 240 km long, had 100 locks, and passed through a 165 m long tunnel. No such engineering feat had been attempted in Western Europe since the fall of the Roman Empire, and it was, in Voltaire's words, 'the most glorious moment' of the reign of Louis XIV. It pointed the way to a new era in the peaceful use of gunpowder, which was to last for almost 200 years.

At first, before the railways and the roads, the emphasis was on canals and in Britain two generations of canal builders, of whom Brindley, Rennie and Telford are the most famous, began their work. In 1757 the Sankey Navigation linked the St Helens' coal-fields to the River Mersey and gave an outlet to the sea through Liverpool; in 1759 the Bridgewater Canal, with a tunnel at both ends and an aqueduct carried on arches across a valley, was built to carry coal from the Worsley mine, belonging to the third Duke of Bridgewater, into Manchester; the Caledonian Canal was begun in 1804; a 2.7 km tunnel was built under the Pennines in 1811 to carry the Leeds and Liverpool Canal; and the Regents Canal was constructed in 1826. By 1858, when the system was at its height, there were 7,000 km of canal covering most parts of central England and travelling through 75 km of tunnel. Many rivers had also been canalized to make them navigable.

Across the Atlantic, the Erie Canal, 585 km long, was begun in 1817 and finished eight years later. This linked the mid-western farmlands of Albany with New York and began the growth of the latter into such a major port and city; the Welland Canal, by-passing Niagara Falls, was constructed in 1827; and the first canal tunnel was built in Pennsylvania in 1828. There was such frenzied activity that local gunpowder manufacturers could not keep up with the demand for their product.

But the extremely profitable boom in canal usage eventually gave way to the rise in rail transport, which enhanced still further the progress of industrialization begun by the canals, and finally made the existence of large towns possible. In 1804 Trevithick built a steam locomotive that pulled a load of 10 tonnes at 8 km an hour along a 16 km length of track in Wales, and the Stockton–Darlington railway, using Stephenson's first locomotive, opened in 1825. The 47 km long Liverpool–Manchester railway, leaving Liverpool through a 1.8 km long tunnel, was opened with great ceremony by the Duke of Wellington, and the sensational running down of poor Mr Huskisson by the 9 horse-power, 48 km per hour *Rocket* aroused special interest. By 1869 the Union Pacific and Central Pacific railroads had been joined at Promontory Point in Utah to complete the trans-continental link.

The boom in building railways lasted until around 1880, and what gunpowder had done for the canal it then did for the railroad, though at first it was not required in large quantities because the new tracks were laid on level ground. Later, however, deep cuttings and long tunnels had to be created to carry the ever-increasing traffic. In England Isambard Kingdom Brunel began work on that 'monstrous and extraordinary, most dangerous and impracticable tunnel at Box',[11] which was to complete the Great Western Railway line from London to Bristol. Work began in 1836 and the tunnel, almost 3.3 km long, was opened in 1841. It had cost over a hundred lives. Part of it ran through clay and soft rock and had to be lined with thirty million bricks, but an 800 m stretch to the east passed through hard Bath stone. This part alone took two and a half years to finish and required a tonne of gunpowder for blasting and a tonne of candles for illumination each week. The 12.9 km long Mont Cenis Tunnel under the Alps, built by a Sardinian, Germain Sommeiller, was a still grander scheme. Drilling began in 1857, from both the Italian and French sides, but progress amounted to an average of less than 25 cm a day, at which rate the tunnel would have taken about seventy-five years to finish. Fortunately the rate of advance increased more than ten-fold when pneumatic drills were introduced in 1861, and the two halves were successfully joined on Christmas Day in 1870. Gunpowder was the only explosive that was used.

Beyond the transport scene, gunpowder and, later, other explosives were used in opening up oil wells, in building deep harbours and spacious airports, in demolition work that could remodel acres of landscape, in constructing sewage,

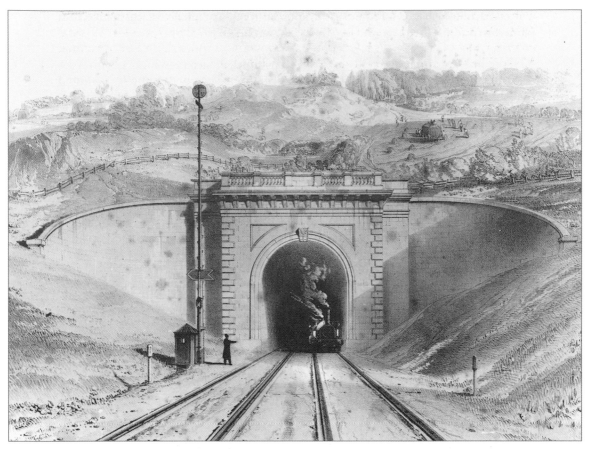

The entrance to the Box tunnel. (National Railway Museum, York)

drainage and irrigation systems, and in the erection of aqueducts, lighthouses, skyscrapers and dams. All the wildest dreams of the civil engineer could be realized but not without the blasting power of gunpowder and its more powerful successors.

*   *   *

So was gunpowder Good or Bad? Take your pick. But neither is the real answer. Gunpowder can destroy or create, and the blame or the credit for what it achieves rests on the shoulders of the human beings who make use of its enormous power. Thomas Carlyle listed gunpowder, along with printing and the Protestant religion, as 'one of the three great elements of modern civilisation'. William Cobbett linked it with banknotes as 'one of the two most damnable inventions that ever sprang from the minds of men under the influence of the devil'.

Let Alfred Nobel, who knew a thing or two about explosives and who had a happy turn of phrase, have the last word. He wrote, in 1875:

That old mixture possesses a truly admirable elasticity which permits its adaptation to purposes of the most varied nature. Thus, in a mine it is wanted to blast without propelling; in a gun to propel without blasting; in a shell it serves both purposes combined; in a fuse, as in fireworks, it burns quite slowly without exploding. Its pressure, exercised in these numerous operations, varies between (more or less) 1 ounce to the square inch [4.4 g per sq. cm] in a fuse and 85,000 pounds to the square inch [5,977 kg per sq. cm] in a shell. But like a servant of all work, it lacks perfection in each department, and modern science, armed with better tools, is gradually encroaching on its old domain.[12]

So, in the end, after something like 600 years in the western world, it was 'Goodbye to Gunpowder' as a major explosive.

CHAPTER 7
Gunpowder Modifications

So attractive was the commercial success awaiting any successful inventors that great ingenuity, and some chicanery, were applied in trying to make improved or special types of gunpowder to meet the ever more demanding requirements of the military, sportsmen, miners, civil engineers and fireworks and fuse manufacturers. The main ends were more power, greater safety, lower cost, a lower burning temperature and less smoke. Some optimists even hoped for noiseless powder, and Peter Whitehorne wrote: 'There be many who bring up lies, saying that they can tell how to make powder that shooting gunnes shall make no noise, the which is impossible.'[1]

Many of the attempts were not only very optimistic, they were also fraught with danger and ended in disaster, for it is unwise to meddle with explosives, and before the Explosives Act of 1875 almost anyone could make and market any sort of mixture. A selected list of contemporary trade names – Earthquake powder, Elephant brand, Fortis, Carbonite, Kinetite, Dynammon, Pembrite, Fractorite and Electronite – gives some idea of the range of invention and the rather bewildering array of products that manufacturers were trying to sell. Alas, the menu was not drawn up without some over-zealous experimenters losing their own lives and contributing to the loss of others.

The major advances came when potassium nitrate was replaced by other oxygen-carriers such as sodium nitrate, potassium chlorate, potassium perchlorate and ammonium nitrate. Smaller changes were made by altering the composition or the method of manufacture of the gunpowder, or by incorporating extra additives to achieve some particular purpose.

* * *

When sodium nitrate first became available from Chile in around 1825, it was thought that it had to be converted into potassium nitrate before it could be used in gunpowder. This was because it absorbed moisture from the air so that any powder that contained it could not be kept dry. Lammot du Pont, overcame the problem in 1857 by limiting the absorption of moisture in a gunpowder made from sodium nitrate by heavy glazing, which gave a protective coating. A considerable amount of graphite was added and the powder was tumbled in a barrel for twelve hours. He called the product 'B' or 'soda' powder to distinguish it from the 'A' powder made using potassium nitrate. The 'B' powder was not

The seizure of the British mail-ship Trent *by USS* San Jacinto *on 8 November 1861. (From* The Illustrated London News, *November 1861)*

suitable for use in firearms, and it did not store as well as the 'A' material. But it was satisfactory for many mining and constructional applications, particularly in hot climates, and it was cheaper than the 'A' powder. It was produced mainly in the United States, enabling Americans to depend less on supplies of nitre from India, which were largely controlled by British agents.

There were times, however, when those supplies were absolutely essential. The Northern States in the Civil War did not at first expect to suffer a shortage of gunpowder, but as the war dragged on, and became fiercer, they began to realize that there was likely to be a scarcity of nitre, which was necessary to make good quality military powder. Lammot du Pont was therefore chosen by Lincoln to go to England, in November 1861, to buy as much nitre as he could lay his hands on. He carried with him $500,000 in gold, and within a few days he had purchased almost two thousand tonnes which was loaded on to ships ready for transportation.

At that moment, a crisis erupted when a British mail-ship *Trent* was seized off Havana by a Northern ship, the USS *San Jacinto*. By chance, two Southern diplomats, John Slidell and James M. Mason, were on board the *Trent* travelling to England to present their side's case. Arrested by the commander of the *San Jacinto*, they were sent ashore as prisoners. What appeared to the British to be piracy at sea led to some demands that they should join the Civil War on the side of the South against the North, but, within a few rather threatening weeks, the incident had been resolved. Britain sent some Guards to Canada, placed an embargo on the exportation of saltpetre and issued a firm ultimatum: Lammot du Pont made it clear to the Northern government that they would get no saltpetre if the matter was not cleared up; and Lincoln ordered the release of Slidell and James early in 1862. The Indian saltpetre was unloaded in the United States shortly after. It is doubtful whether the war could have gone on much longer without it.

*   *   *

About eighty years earlier a Frenchman, Count Claude L. Berthollet, was experimenting with potassium chlorate as a substitute for potassium nitrate in gunpowder. He was born of poor parents, on 9 December 1748, in the Haute Savoie, then part of Italy. After gaining a medical degree at the University of Turin in 1768, he went to work in Paris where he collaborated with Lavoisier in devising a system of chemical nomenclature and laying the foundations of chemistry as a separate science. In 1781 he was elected to the Académie des Sciences, and in 1794 to a professorship at the École Normale. Unlike Lavoisier and Pierre-Samuel du Pont he remained on good terms with the revolutionaries, and Napoleon, with whom he became friendly, partly through teaching him some chemistry, made him a Senator and a Count. But later Claude voted to depose Napoleon and sided with the returning Bourbons before he died in 1822.

Potassium chlorate is a white solid that looks very much like potassium nitrate. It contains a lower percentage of oxygen than the nitrate but is less stable, so that what oxygen it does contain is released at a lower temperature, so much so that simple mixtures using potassium chlorate are too sensitive and do not obey the fundamental rule that an explosive must go off when you want it to, but not when you don't. Sodium chlorate, readily available as a weed-killer, has frequently been the lethal component of amateur home-made explosives. It looks innocuous but it must be handled with extreme caution.

Berthollet had prepared and investigated potassium chlorate in 1786, and he tried it in gunpowder in 1788. To celebrate this innovation a party of guests, including M. amd Mme Lavoisier, were invited to see the first batch of the new powder being made. While it was being pounded in a stamp mill, the party went off to have breakfast, but on their return they were met by a violent explosion which threw the two leading members a great distance and killed them. But the lesson was not learnt everywhere and many other attempts to use potassium chlorate were to follow.

Kellow and Short, who had at one time been associated with the slate quarry at Delabole in Cornwall, made a particularly bold attempt in 1862. They added sulphur, tan and sawdust to a hot solution containing, for good measure, potassium nitrate, sodium nitrate and potassium chlorate, until the mixture formed a paste which was spread out on trays to dry. It was claimed that this product was four times as effective as ordinary gunpowder when used for blasting in mines, that it produced only half as much smoke, that it was less affected by damp, and that it was cheaper. They called their process the 'Anti-Explosion' method of manufacture. It was too good to be true, and they were better salesmen than chemists. During the next three years, there were five or six explosions at the works, and on Wednesday 5 July 1865 a devastating blast demolished an entire block of buildings, ending another chlorate adventure. Fortunately, and miraculously, no lives were lost.

Potassium chlorate was eventually used successfully in explosive mixtures. Nobel wrote to a friend: 'Your fear of chlorate of potassium is exaggerated. When it smells of sulphur it is as sensitive as an hysterical girl, and when it feels phosphorus on its surface it is worse than a thousand devils. But it can very well be tamed down to keep itself within the nurture and admonition of the Lord.'[2] The taming was achieved by incorporating castor oil or rosin or paraffin wax into explosive mixtures containing potassium chlorate. Cheddite, for example, developed at Chedde in France, was a successful mining explosive which contained potassium chlorate and castor oil together with nitronaphthalene. Steelites, invented by Everard Steele of Chester, England, contained potassium chlorate with specially treated rosin, and similar explosives in Germany were called Silesia mixtures. Sodium chlorate, potassium perchlorate and ammonium perchlorate were also used satisfactorily.

* * *

Not all the explosive mixtures were as dangerous as the early potassium chlorate ones, but the combination of several factors, including the fact that almost anyone could make and sell them, the continuing numbers of accidents (culminating in the deaths of fifty-three people in an explosion at Messrs Ludlows' in Birmingham), and the advent of new types of explosive other than gunpowder, led to the passing of the Explosives Act in 1875, which was extended in 1923. The Act laid down a set of rules and regulations which, together with the issue of later Orders in Council under the Act, have controlled the explosives industry ever since. Its success was shown by the rapid fall in the number of deaths annually in the industry in Great Britain from around thirty-two in 1875 to seven by the turn of the century. This, too, was at a time when new explosives were coming into use and when the overall demand was continually rising.

The Act stipulated the conditions under which explosives had to be manufactured, handled, transported and stored, and a team of specialists – Her Majesty's Inspectors of Explosives – was set up, under the control of the Home Office, to monitor the situation. Colonel Sir Vivian D. Majendie, who had played a big part in drafting the Act, was the first Chief Inspector, and he set a very high standard which has been maintained ever since, to the benefit of all concerned. The inspectors were empowered to enter any establishment to check on its practices, they investigated every accident, and they issued annual reports.

Any new home product, as well as anything imported from overseas, had to be submitted for testing before it could be licensed for use and the Home Office issued a list of authorized explosives. Today's list is produced by the Explosives Inspectorate of the Health and Safety Executive. Amendments are published

HSE
Health & Safety
Executive

Explosives Acts, 1875 and 1923
The Classification and Labelling of
Explosives Regulations 1983 (CLER)

LIST OF CLASSIFIED AND
AUTHORISED EXPLOSIVES
1994
(LOCAE)

The title page of the 1994 LOCAE publication, an alphabetical list with over six thousand entries. (Crown copyright, reproduced with the permission of the Controller of Her Majesty's Stationery Office)

annually, and a new edition is prepared every three years. The list contains over six thousand named items covering the whole range of explosive products. There are, for example, 6 types of Bengal sparkler, 1 joke cigarette, 366 named cartridges of explosive, 430 detonators, 120 seat ejector rockets, 20 sorts of cordite, 39 types of RDX, 6 bird scarers and 157 oil well cartridges. A corresponding list of military explosives is prepared, as a restricted document, by the Explosives Storage and Transport Committee of the Ministry of Defence (ESTC).

There were particular safety problems in the use of explosives in coal-mining and the Coal Mines Act of 1887 laid down that only explosives that had passed a series of special tests could be used. They were called safety or permitted explosives in Great Britain, and permissible explosives in the United States. They had to be specially labelled, and the quantity of explosive that could be used at any one time had to be specified. The first official list of such explosives was isssued in Great Britain in 1899.

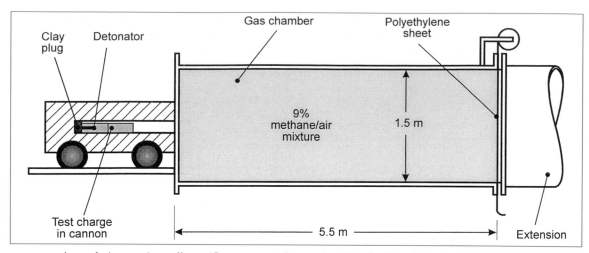

An explosives testing gallery. (Crown copyright: produced by the Visual Presentation Services of the Health and Safety Laboratory)

The special tests involved the use of the Trauzl test and later the ballistic pendulum to measure the power of the explosive, and also a test to examine its effect on the sort of gas mixture that might be present in a coal-mine. The latter test involved firing measured quantities of the explosive from a gun into a mixture of air containing coal–gas, methane or coal–dust.

The first testing gallery was built at Woolwich in 1897 and a larger one at Rotherham in 1911, but since 1929 the tests have been carried out at Buxton by what is now the Explosives Section of the Health and Safety Laboratory. A cannon with a bore 50 mm in diameter (to simulate a borehole in a mine) is used for containing the explosive under test. The cannon is positioned and sealed into a hole at one end of a steel cylinder, which is intended to represent an underground roadway in a mine. The cylinder is 15 m long, has a diameter of 1.5 m and is open at the end remote from the gun. A plastic diaphragm is fitted inside the cylinder 5.5 m from the mouth of the gun and the enclosed space at that end of the cylinder is filled with a mixture of methane and air, or coal–dust and air, similar to that likely to be found in coal-mines. Known quantities of explosive are then fired in the cannon, and in this way the maximum charge that can be used without exploding the gas mixture in the cylinder can be accurately measured. The detailed conditions under which the test can be carried out can be varied in a number of ways to simulate different practical situations, and nowadays five classes of permitted explosive are recognized in Great Britain, each one designed for a particular purpose and tested under relevant conditions and against suitable criteria. In official lists of explosives, they are labelled with a P.

* * *

It is now known that an explosive with a long-lasting, hot flame is much more likely to ignite fire-damp than one with a cooler flame of shorter duration because it is necessary to raise a methane-air mixture above its ignition temperature for a definite length of time before it will explode. And it was appreciated, before any legislation was enacted, that it was necessary to lower the burning temperature of gunpowder if it were to be used safely in coal-mines.

Reducing the sulphur content lowers the burning temperature, but the same result was achieved more effectively by incorporating certain additives. Bobbinite, for example, was a popular permitted explosive in coal-mines for many years. It consisted originally of 64 parts of nitre, 2 of sulphur, 19 of charcoal and 15 of a mixture of ammonium and copper sulphates. Later the sulphates were replaced by a mixture of paraffin wax and starch. The inert additives made Bobbinite weaker than gunpowder but this had some advantages in blasting a soft material such as coal when it was required to displace it without too much fragmentation. Bobbinite eventually passed the Woolwich test, which normal gunpowder could not do, but failed the more stringent test applied at Rotherham. After 1913 it was only allowed to be used in non-fiery coal-mines and even then only for a limited period of five years. So, like many other modified gunpowders, it had its day and ceased to be. But it is still common to add so-called fire-depressants, such as salt, to explosives used in coal-mines to lower the temperature.

A group of explosive mixtures containing ammonium nitrate had a much longer life than Bobbinite. They were introduced by two Swedes, Ohlssen and Norrbin, in 1867, but it was a Frenchman, M. Favier, who first realized that they were safer to use in coal-mines than gunpowder. Austrian Dynammon contained ammonium nitrate and charcoal; Cornish Fumelssite contained ammonium nitrate mixed with sawdust; Pembrite contained ammonium and barium nitrates with vegetable oil and sulphur. These were the forerunners of the ammonals patented in Vienna in 1900 and containing originally ammonium nitrate, charcoal and powdered aluminium. They were widely used in civilian mining and later, when a little TNT was added, in quarrying; they were also used in land-mines in both the First and Second World Wars.

Ammonium nitrate is quite cheap but it absorbs moisture from the air, like sodium nitrate, so explosives containing it had to be well sealed in waterproof wrappings. For some years, too, it was difficult to store it safely. If it became moist and then dried out it caked into a hard mass and in this state it was somewhat unstable. At Oppau in Germany in 1921, 4,500 tonnes of such a mass, mixed with ammonium sulphate, was being broken up by blasting when it all detonated. Over 500 people were killed, almost 2,000 were injured, and buildings were demolished up to 6.5 km away. Again, in Texas City in 1947 a large load of ammonium nitrate blew up in a ship in the harbour after a fire on board.

Fortunately, these problems have been overcome by new methods of manufacture and ammonium nitrate is nowadays the major ingredient in a host of commercial explosives because it is the cheapest suitable oxygen-carrier.

* * *

Despite the vast number of different products which were put on the market in the nineteenth century the use of gunpowder and gunpowder-type explosives began to decline towards the end of the century as nitro-explosives such as dynamite and guncotton became available. But a series of patents taken out in 1871 by Dr Hermann Sprengel, the inventor of the mercury vacuum pump, together with the introduction of ammonium nitrate, provided a fillip and eventually a resurgence in the use of gunpowder-type mixtures, at least in mining.

Sprengel's idea was that two substances – one an oxygen-carrier and the other a fuel – should be mixed on-site as and when required. The advantage of this was that neither chemical need be an explosive on its own so that it could be transported and stored without difficulty. Potassium chlorate, manganese dioxide and nitric acid were listed as possible oxygen-carriers; nitrobenzene, turpentine, naphthalene and paraffin wax as fuels. Rack-a-rock, a popular Sprengel explosive, consisted of cartridges of potassium chlorate in cotton bags which were dipped into nitrobenzene until they had absorbed between a quarter and a third of their weight. Over 100 tonnes of this explosive were used in 1885 to break up the Hell Gate reefs at the mouth of the East River in New York harbour. It was also used in the Russo-Japanese war of 1904, and in building the early Chinese railways. Sprengel-type explosives were much favoured in America and the Far East but other countries, including Great Britain, discouraged their use, regarding the on-site mixing operation as an 'explosive manufacturing process' requiring a licence, which was very difficult to obtain.

The next step forward was the use of liquid oxygen as the oxygen-carrier in what became known as LOX (Liquid OXygen) explosives. It became possible to make them in around 1900, when Linde devised his method of liquefying air on a commercial scale, and when Sir James Dewar, working in the Royal Institution, invented the Dewar flask in which liquefied gases could be stored. That flask, originally intended for scientific purposes and not patented by its inventor, has become today's Thermos flask. Linde used charcoal, cotton wadding or kieselguhr to absorb his liquid air or oxygen in what were called Oxyliquit explosives, but lampblack was used in later mixtures. The explosives were quite powerful and they were cheap to use, particularly on a large site where the oxygen-producing equipment could be run continuously. They had to be fired shortly after absorption of the oxygen, otherwise it escaped. For that reason, large cartridges which lasted longer were most convenient and this again favoured their use on large sites.

LOX explosives were used in excavating the Simplon tunnel as early as 1899 but their subsequent use in Europe was limited. Their main use was in large opencast coal-mining operations in the United States, particularly between 1940 and 1950. Like other explosives, however, they were not entirely safe and a number of peculiar and inexplicable fatal accidents contributed to their decline. It was simply another case in the general history of explosive mixtures – here today, almost gone tomorrow – though liquid oxygen survives as one of the main ingredients of modern rocket propellants, generally with liquid hydrogen as the fuel, even though the hazards of using it have not been completely eliminated.

The demise of LOX explosives was also accelerated by the introduction, in the

United States in the early 1950s, of Akremite, a Sprengel-type explosive made by mixing carbon black or coal-dust with specially prepared spherical granules of ammonium nitrate to facilitate the mixing process. The product was safe, powerful, easy to use and cheaper than LOX mixtures. A further improvement was made by replacing the carbon black with fuel oil, similar to diesel oil; the product is called ANFO. The mixing can be done on a small scale, as in making concrete, by pouring the fuel oil on to a pile of ammonium nitrate and blending the two with a wooden spade. Alternatively, ANFO can be delivered ready mixed in special trucks or bought in packs. A typical mix contains 94 per cent ammonium nitrate and 6 per cent oil. It can be poured into a dry borehole or blown in under pressure.

The earth-moving ability of ANFO exceeds that of dynamite and it is safer and considerably cheaper. It has two main drawbacks: it is soluble in water so that it cannot be used, in its simple form, in wet conditions unless it has been carefully packaged, and it requires a primer of a conventional explosive, or a special, high-powered cap, for detonation because it is not very sensitive. But its advantages far outweigh these disadvantages and it has been used increasingly since its introduction in the late 1950s.

Many of the disadvantages of ANFO were overcome in the early 1960s by the development of slurries or water-gels that can be used in dry or wet conditions. The base consists of a suspension of ammonium nitrate in water. This is mixed with a sensitizing fuel such as TNT or finely divided aluminium, together with thickening and gelling agents, such as guar gum, to impart water-resistance. The slurry may also be aerated by the introduction of minute, suspended air bubbles, or micro-balloons consisting of hollow spheres of glass or plastic. The products vary from thick soupy liquids to much more rigid gels; they can have densities greater than 1 so that they will sink in water-filled boreholes; they can be of varying power; they can be mixed on-site or pre-packed; and they can be set off by standard detonators.

The slurries were followed in the late 1960s by emulsions. They contain an emulsion of ammonium nitrate solution and oil, with entrapped microscopic air bubbles suspended in surrounding oils and waxes, which serve as the fuel. They have the advantage over slurries of being more water-resistant, more flexible, and quicker and cleaner in use.

The water content of the slurries and emulsions might be expected to make them less powerful for blasting than dry ANFO, but the reduction in power is not, in fact, very great. This arises because the semi-liquid product fills any borehole more completely than a solid mixture can, and moreover the presence of the water causes the production of a higher volume of gas when the product detonates.

Since 1992 ICI have produced a number of very successful products. In their 'Handibulk' system, non-explosive ingredients, which can be stored and transported safely, are delivered in a specially designed truck to a site where they can be mixed. Alternatively, the emulsion explosives are manufactured in pre-packed cylindrical cartridges, weighing between 113 and 12,500 grams, under the trade names of Powergel and Magnum. This new generation of explosives has

An emulsion bulk truck carries safe, non-explosive ingredients which are mixed at the point of use. (ICI)

largely replaced nitroglycerine and TNT products for quarrying and mining, and they are being used in a number of different ways. In air-decking, for example, improved blasting is achieved when air gaps, or air-decks, are inserted in a column of explosives in a drill hole, by using self-inflating gas bags. The technique was pioneered in Russia in the 1970s and then in Australia and the United States. It has been used increasingly in Great Britain since 1995.

Computer-aided blasting programmes, developed in the last ten years mainly in South Africa and Australia, are also coming to the fore. They enable a blasting operation to be carried out in the most efficient way by using a computer

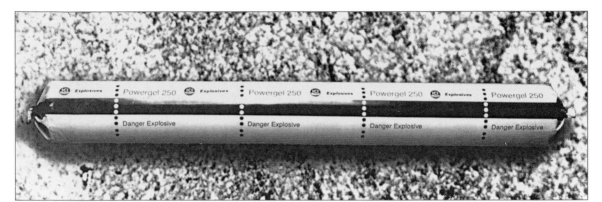

A cartridge of Powergel 250 *packaged emulsion explosive. (ICI)*

programme, exemplified by ICI's SABREX 3, to plan the size and spacing of the drill holes, the type and amount of explosive required in relation to the nature of the rock face, and the timing of the firing of each drill hole. It is highly sophisticated and can be very successful compared with the older, rather hit-or-miss approach.

The seeds that Sprengel planted in 1871 have borne much fruit. The crystal ball, as yet, reveals nothing of the future. What might it possibly be? Can the basic gunpowder formula be modified still further? Why not? Many might have thought that the formula was worked out by 1890 but there is still some magic in it a hundred years later.

CHAPTER 8

Nitroglycerine

Angina pectoris is a disease of the heart marked by paroxysms of intense pain, which until 1879 could only be treated by the temporary palliatives of brandy, opium, ether, chloroform or bleeding. That nitroglycerine – the main ingredient of dynamite, blasting gelatine and gelignite – came to be used as a treatment is perhaps one of the oddest examples of the diverse uses to which a single chemical can be put.

Nitroglycerine (NG) was first made by an Italian, Ascanio Sobrero, who graduated from the University of Turin in 1832 with a degree in medicine. When circumstances forced him to abandon his chosen career he turned to chemistry, studying under Theophile Pelouze in France and Justus von Liebig in Germany. Back at his old university as Professor of Applied Chemistry, he made nitroglycerine in 1847 by treating glycerine with a mixture of concentrated sulphuric and nitric acids. He described it as a yellowish oil; discovered, when a small sample blew up and spattered glass fragments into his hands and face, that it was violently explosive; and found, by tasting, that it had unexpected physiological effects. He reported:

> It has a sharp sweet, aromatic taste. It is advisable to take great care in testing this property. A trace of nitroglycerine placed upon the tongue, but not swallowed, gives rise to a most pulsating, violent headache, accompanied by great weakness of the limbs. A dog was given a few centigrams of nitroglycerine. It soon began to foam at the mouth and then vomited. Despite the fact that the greater part of the nitroglycerine had thus been eliminated from the system, within seven or eight minutes the animal had fallen down and almost ceased breathing. A dose of olive oil and ammonia was administered. The animal revived somewhat, and remained for some two hours whining, trembling violently, and beating its head on the wall. A post-mortem examination showed that the vessels of the brain and heart were suffused with blood and much distended. Similar results were obtained with rats and guinea-pigs. The violence of the decomposition has prevented me, up to the present, making an analysis of the body.[1]

Sobrero realized that he had made a very unusual substance, but he did little or nothing to exploit its potential. It was his aim 'that science should not be made a pretext or means of dishonourable undertakings or of business speculations',[2] and his memory is perpetuated by a statue in Turin.

The statue of Ascanio Sobrero in Turin. The inscription reads:

Illustrious Chemist
Discovered explosives of extraordinary power
Professor Emeritus
He stimulated the progress of industry
Under the auspices
of the Turin Association
of Industrial Chemistry

(Giuseppe Antonioli: Italesplosivi, Milan)

The first medical trials of nitroglycerine were in the field of homeopathy, a system of medicine pioneered by the German, Samuel Hahnemann (1755–1843), and based on the principle that like cures like – *similia similibus curantur*. It followed that nitroglycerine, which certainly causes headaches, would cure them. Against this background, it was tried out by Drs Hering, Davies and Jeanes at the Hahnemann Medical School in Philadelphia and recommended for treating a number of diseases.

A British doctor, A.G. Field, used it in 1858 in more orthodox allopathic treatment of headaches, toothache and neuralgia, but it was left to Sir Thomas Lauder Brunton, of the Royal Infirmary in Edinburgh, to establish nitroglycerine as an effective treatment for angina pectoris. He was himself a sufferer of the disease and he had discovered that amyl nitrite was a good drug to take. When it was found that amyl nitrite and nitroglycerine were similar they were both used for treatment, and the latter is still in use today. It causes dilation of the blood vessels.

One method of treatment supplies the daily dose of nitroglycerine in a sealed self-adhesive patch which is stuck on the patient's skin. This enables the drug to diffuse into the body so that the dosage is continuous and prolonged for twenty-four hours, but care is necessary for there has been a report of the patch exploding with a sharp, but harmless, bang when a patient was subjected to electric shock treatment. Too high a dosage of absorbed nitroglycerine causes severe headaches, and, when it came to be widely used as an explosive, workers who handled it continually were always liable to suffer from what they called 'NG head'. Most workers became acclimatized to it after a few weeks, but if they were away from their work for a while, they had to go through the miserable period of acclimatization once again. To avoid this, many of them used to carry a small amount of the explosive around with them, despite all the rules and regulations against it.

* * *

The results of the first experiments to test the explosive characteristics of nitroglycerine, carried out in Italy, were not very promising, because the oily liquid proved to be very capricious. At times it would explode quite unexpectedly; at others it was difficult to detonate. The credit for bringing it under control lies with two Swedes, Immanuel Nobel and his son Alfred, and to a lesser extent two Russians, Nikolai Zinin, a professor of chemistry and Vasilii Petrushevskii, a chemist and artillery officer.

Alfred Nobel, who plays the main part, was fourteen years old when Sobrero discovered nitroglycerine. At the time, Alfred was living in St Petersburg with his father, mother and brothers, and by coincidence was being taught chemistry by Professor Zinin. By the time he was thirty he was back in Sweden helping his father to manufacture nitroglycerine on a small scale and making the first of his epoch-making discoveries, patented in Britain in 1863. He had found that the most reliable way of setting off nitroglycerine was to use the explosion from a small amount of gunpowder. Explosives such as gunpowder had previously been set off by fire or fuse but these would not ignite nitroglycerine satisfactorily. It was the idea of using one explosive to detonate another that was new. Like so many of the best inventions it was simple – so simple, perhaps, that anyone could have thought of it. But anyone didn't – Nobel did. And that was just the start of the new age of high explosives, which he was to dominate. In 1867 he patented dynamite, in 1875 blasting gelatine, and in 1888 ballistite. Before he died in 1896, aged sixty-three, he had registered 123 patents in Great Britain alone and well over 300 worldwide, and nitroglycerine was being manufactured all over the world from glycerine, nitric acid and sulphuric acid.

* * *

The glycerine part of nitroglycerine originates from the ancient art of soap-making, which is almost two thousand years old. Originally oils and fats, such as olive oil and mutton fat, were boiled with wood ashes. Later the range of oils and fats was extended and the wood ashes, for which the Arabic name was *al kali*, were replaced by other alkalis such as quicklime, sodium hydroxide or potassium hydroxide. The soap forms as a solid, which can be filtered off; the filtrate was originally known as 'spent lye' or 'sweet water' because of its taste. Karl Wilhelm

Scheele, a compatriot of Nobel's, first showed in 1799 that the sweet taste was due to the presence of glycerine; he called it 'oelsuss' or 'sweet principle of oils'. Whereas this by-product used to be discarded as useless, it is today regarded as almost as important as the soap itself.

Scheele, working mainly on his own as a poor apothecary, proved to be one of the world's all-time greats in chemical discovery. In his short life – he was only forty-four years old when he died – he discovered not only glycerine but also oxygen, ammonia, hydrogen chloride, milk-sugar, and hydrofluoric, molybdic, tungstic, lactic, gallic, citric, tartaric, pyrogallic, uric and malic acids. In his own words: 'to explain new phenomena, that is my care, and how glad an investigator is when he finds what he has sought so industriously; it is a pleasure that fills the heart with joy. For it is only the truth that we wish to know, and what joy to have discovered it.'[3] Another scribe chose the appropriate phrase 'Avec de petite ressources, il fit de grandes choses.'[4]

Pelouze, Sobrero's early teacher, worked out the formula of glycerine in 1836, and it was eventually recognized as a type of alcohol; this is why, in chemistry terms, the name glycerol is preferred to glycerine. Conversion into nitroglycerine simply involves warming the glycerol with a mixture of concentrated nitric and sulphuric acids. This replaces three OH groups in the glycerol molecule by three NO_3 (nitrate) groups, which is why the name glyceryl trinitrate is a more systematic, though less common, name for the product.

*   *   *

It is probable that both sulphuric and nitric acids were known to Jabir ibn Hayyan, or Geber, a great Islamic alchemist who lived around AD 800, although there are some doubts as many later writers tried to authenticate their works by attributing them to Geber. The acids were certainly being made and used, both as solvents and in medicine, in the mid-sixteenth century. Their present names and formulae date from around 1790.

Sulphuric acid was first made by heating sulphates, particularly ferrous sulphate, which, because its crystals look like bits of green glass, was called green vitriol. The fumes given off when it was heated were passed into water to form what was called oil of vitriol, now known to be sulphuric acid. Basil Valentine, a monk with a strong interest in chemistry and medicine, described the process in his *Last Will and Testament* (1656): 'If you get such deep graduated and well prepared Mineral, called Vitriol, then pray to God for understanding and wisdom for your intention, and after you have calcined it, put it into a well coated retort, drive it gently at first, then increase the fire, there comes in the form of a white spirit of vitriol in the manner of a horrid fume.'[5]

By 1740 the acid was being manufactured on a small scale by absorbing the fumes from a burning mixture of sulphur and saltpetre (shades of gunpowder) in water. The process was carried out in glass jars with a capacity of about 300 litres. The first factory was built in Twickenham by Joshua Ward, a quack doctor, who wanted the acid for medical purposes. Local complaints against the pollution that it caused forced the factory to move to the less densely populated area of

Richmond: even in those dark days environmentalists were not inactive. In 1756 John Roebuck replaced the glass vessels by lead boxes so that the scale of the operation could be increased and, with suitable modifications, this soon broadened out into the lead-chamber process which was used all over the world for making sulphuric acid until it began to decline after 1900 with the advent of the modern Contact process.

In this process sulphur or metallic sulphides are first heated in air to produce sulphur dioxide gas. This is then mixed with oxygen and passed over a hot catalyst to form sulphur trioxide, and subsequent absorption of this in water forms sulphuric acid. The process was patented in 1831 by a Bristol vinegar manufacturer, Peregrine Phillips, but it was not operated until about 1900 because the lead-chamber acid satisfactorily met all the requirements until then. At that time, however, the expanding dyestuffs industry required stronger acid than could be made by the lead-chamber process, and later on the First World War proved that the manufacture of explosives such as TNT also needed purer acid. By about 1960 all the world's sulphuric acid was being made by the Contact process and the lead-chamber process was obsolete.

The vast importance of sulphuric acid to the chemical industry was emphasized by Lord Beaconsfield's famous remark that 'there was no better barometer to show the state of an industrial nation than the figure representing the consumption of sulphuric acid per head of population', and by the First World War slogan 'Sulphuric acid, the life-blood of chemical industry'. These sentiments are less true today than they were seventy years ago, but sulphuric acid is still among the top ten most important chemicals.

*   *   *

Nitric acid was first made in the middle of the sixteenth century by heating green vitriol with alum and saltpetre. The liquid product was called 'aqua fortis' because it was powerful enough to dissolve all known metals except gold. By adding salt or sal-ammoniac a still more powerful mixture could be made which would dissolve gold; this was called 'aqua regia'. The use of nitric acid for separating silver from gold led to it being called *Scheidwasser* ('the separating water') in Germany. It was also known as spirit of nitre.

In 1648 Glauber discovered that heating saltpetre with sulphuric acid was a better method of making nitric acid, and this method was universally used by 1800. With the eventual replacement of saltpetre by Chile saltpetre, when it became available, the process was operated until about 1925. It had begun to be replaced much earlier than that as a result of some remarkable research carried out in Germany by Fritz Haber at the start of the twentieth century. Haber, born in 1868, was the son of a prosperous Jewish chemical manufacturer and merchant, and after a thorough chemical education in Berlin, Heidelberg, Charlottenburg and Karlsruhe, it had been intended that he should join the family firm. But after some months in his father's factories, he decided that he preferred an academic to an industrial career, and he soon became Professor of Chemistry at the Technische Hochschule in Karlsruhe. It was there that he established a laboratory

Fritz Haber. (BASF)

which became the world leader in high-pressure research and where he developed what is now known as the Haber process for making ammonia.

Sir William Crookes had pointed out in 1898, in his presidential address to the British Association for the Advancement of Science, that the natural supplies of nitrogen-containing compounds, so essential for use as fertilizers, showed signs of running out as the world population grew. 'The fixation of nitrogen', he said, 'is vital to the progress of civilised humanity.'[6] This gave a great impetus to research into methods of converting the vast amounts of nitrogen gas available in the atmosphere into nitrogen compounds, which is what the Haber process succeeded in doing. The raw materials of the process are, basically, air and water or methane, all ideal as they are cheap and plentiful. Nitrogen is obtained from the air, and hydrogen from the water or methane, and the two gases are made to react at a temperature of 550°C, under a pressure of between 150 and 300 atmospheres, and in the presence of a catalyst, to form ammonia gas. The process was something of a landmark, first because the conditions required to obtain a good yield of ammonia were worked out from first principles, and secondly because it was a breakthrough in high-pressure technology. It was such a breakthrough, in fact, that it was at first regarded with much incredulity and suspicion by many industrial chemists. Fortunately its merits were recognized in 1909 by Dr Carl

Bosch, the chief engineer of the large chemical firm Badische Anilin und Soda Fabrick (BASF). Within three years he had adapted the process for operation on a large scale and built the first factory at Oppau. At the same time Haber had moved to become Director at the newly established and privately endowed Kaiser Wilhelm Institute which was opened in 1911.

The work of Haber and Bosch was singularly timely for Germany, for what had been started with a view to increasing fertilizer production was soon to ensure adequate production of explosives. At the start of the First World War, Germany was still quite heavily reliant on supplies of saltpetre from Chile for her nitric acid requirements. It is possible that the war may have been over within a few months had it not been for the discovery of 20,000 tonnes of Chile saltpetre in the docks at Antwerp which kept things going. Further supplies were denied to the Germans by the British naval blockade, but no time was lost in manufacturing more and more ammonia and in perfecting a process for converting it into nitric acid by mixing it with oxygen and passing it over a hot catalyst. By the end of the war this had proved to be such an efficient way of making both ammonia and nitric acid that it was adopted worldwide as new plants were built.

During the war Haber had replaced his academic gown with a military uniform when, at the age of forty-six, he was promoted from NCO in the Reserve to Captain and put in charge of all Germany's gas warfare and gas defences. Many of the activities in which he was involved were extremely contentious and appeared to infringe existing agreements, but he was a deeply patriotic, even chauvinistic, Prussian not inclined to deny the State's wisdom or decisions and he had no doubt that he was doing his duty. He wrote: 'A man belongs to the world in times of peace, but to his country in times of war.'[7] But his wife was less happy with his activities and many thought that they were the main reason for her suicide in 1916.

By the end of the war Haber was worn out, dejected and despondent. Even the award of the Nobel Prize for Chemistry in 1919, which might have been expected to cheer him up, was somewhat marred when some British, French and American scientists denounced it because of his wartime activities. Haber had exerted great influence both before and during the war and his life after it was not uneventful. In order to help to pay some of the reparations imposed on his country by the Treaty of Versailles, he attempted to make money by trying to develop a commercially viable method of extracting gold from sea-water. Alas, he found, like others before and since, that gold was obtainable from sea-water only at a price that rapidly led to bankruptcy. So his plans collapsed in 1928, with Haber's pride and reputation distinctly hurt.

Worse was to come. All his strenuous efforts to re-establish the international standing of the Kaiser Wilhelm Institute were shattered in 1933 when he was ordered by the Minister of Art, Science and Popular Education to sack all his Jewish workers. This was too much for him. He resigned, with great dignity, writing to the Minister, 'My tradition requires of me in a scientific post that in choosing fellow workers I take into account only the personal qualifications and the character of the applicants without asking about their racial disposition.'[8] The great chemist, soldier and patriot was now himself just another isolated Jew. He suffered from heart trouble and, after a brief period working in the Cavendish Laboratory in Cambridge, England, he moved to Switzerland where, broken in health and spirit, he died in 1934.

By 1952 many attitudes had changed and a memorial to Haber was erected in the Kaiser Wilhelm Institute. The inscription reads:

> Thermistocles has come down in history not as the exile at the court of the Persian king but as the victor at Salamis. Haber will go down in history as the gifted discoverer of the process of combining nitrogen and hydrogen, which is the basis of the industrial fixing of nitrogen from the atmosphere, who in this way, as was stated when the Nobel Prize was conferred on him 'created an exceedingly important means of improving agriculture and the welfare of mankind, who made bread from the air and scored a triumph in the service of his country and all mankind.[9]

That, one might have thought, would be the end of the Haber story. Not so. At a ceremony at the University of Karlsruhe in 1968 to commemorate the centenary of Haber's birth, two young men suddenly unfurled a banner reading

> Feier für einen Mörder
> Haber = Vater des Gaskriegs[10]

Forgiving and forgetting is not part of everyone's nature.

* * *

Alfred Nobel was born on 21 October 1833 in Stockholm. It was, no doubt, a happy event, even though he suffered poor health, particularly in his early years. As he himself wrote in an autobiographical poem at the age of eighteen:

> My cradle looked a death-bed, and for years
> a Mother watched with ever anxious care,
> so little chance, to save a flickering light.[11]

His mother, Andriette, was to need all her cheerfulness, stamina and strength of character through many of the forty-five years of her notably for-richer-for-poorer marriage. She bore eight children in all but only three, Alfred and his elder brothers Robert and Ludvig, survived to the age of twenty-one. There was a setback immediately after Alfred was born when Immanuel, his father, who had tried to establish himself as an inventor in Sweden, was declared bankrupt. To make a fresh start, he left Sweden to find work first in Finland and then in Russia, leaving his family behind, reliant only on a small milk and vegetable shop in which his wife had been installed by some friends. Alfred, it is said, contributed to the family income by selling matches in the street at the age of seven.

Immanuel's luck changed in St Petersburg. He had invented a sea- or land-mine, filled with gunpowder, while still in Sweden, but only the Russians had appreciated its potential and they had given him financial assistance to set up an engineering firm – Fondéries & Ateliers Mécaniques Nobel & Fils – which made machine-tools, gun-carriages and heating equipment as well as the mines. The mines were, in fact, the first ever to be moored in a shipping lane, and they did some damage to a squadron of small British and French ships which unknowingly sailed into them in the Baltic in June 1855.

So it was that in 1842 the Nobel family was able to move to Russia. Alfred was then nine and his father was prosperous enough to provide him with an expensive private education. Later, at the age of seventeen, he spent three years travelling in Germany, France, Italy and North America before returning to work with his father in St Petersburg at the start of the Crimean War in 1853. Nobel's firm prospered, particularly during the war, and he was able to pay off all his Swedish creditors. But the end of the hostilities in 1856 brought a collapse of business and Immanuel was, once again, declared bankrupt. Poorer than ever, and with four surviving sons, including their youngest son, Emil, he and his wife returned to Sweden in 1859. Alfred joined them soon after but Robert and Ludvig stayed behind in Russia to re-establish their fortunes. Despite many ups and downs, they did this to such good effect that, within fifteen years, they were able to branch out into the reorganization and exploitation of the crude oil deposits in the Baku area of the Caucasus. But that is another story. Suffice it to say that both this new venture and Immanuel's original engineering firm grew into giant world enterprise. But before that, alas, nothing seemed to go well for very long for the Nobels and when the empire of the Tsars collapsed in 1918 the Nobel empire in Russia collapsed with it. The family had to flee as refugees, leaving everything behind and suffering severe hardship.

Back in Sweden, Alfred and his father were occupied in trying to realize the dreams they had had in Russia, along with Professor Zinin, as to how nitroglycerine could be used. They had set up a small laboratory at Heleneborg on the outskirts of Stockholm using capital which Alfred had managed to raise in Paris. But the problem of how to get a controlled, safe explosion from nitroglycerine remained unsolved. At first they tried using a mixture of 10 per cent nitroglycerine with 90 per cent gunpowder but this was not very successful. Nevertheless, it was probably this mixture that gave Alfred the idea of using gunpowder to detonate nitroglycerine which led to the invention of his patent detonator.[12]

In the first design gunpowder was packed into a glass or wooden tube which could be immersed in the nitroglycerine. The gunpowder could be ignited by a fuse and its explosion detonated the nitroglycerine. In Nobel's own words: '. . . the real era of nitroglycerine opened with the year 1864 when a charge of pure nitroglycerine was first set off by means of a minute charge of gunpowder'.[13] Three years later the detonator was improved by substituting mercury fulminate for gunpowder. It was not only the new era of nitroglycerine – it was the start of high explosives.

When nitroglycerine explodes it produces a large volume of hot gases and a consequent big increase in pressure, just as gunpowder does, but the effect is on an altogether different scale. Whereas the rapid burning of gunpowder produces pressures of up to 6,000 atmospheres in a matter of milliseconds, the

Alfred Nobel and an early trade mark. (Crown copyright: Royal Commission on the Ancient and Historical Monuments of Scotland)

The old method of making nitroglycerine. The operator sat on a one-legged stool so that he would not fall asleep and let the mixture overheat. (ICI)

decomposition of nitroglycerine needs only microseconds and can give rise to pressures up to 275,000 atmospheres. This is why gunpowder is a low, and nitroglycerine a high explosive. It is the difference between being bumped into by a pedal cyclist or being knocked for six by an express train. So different is the effect that the word detonating (from the Latin *tonare*, to thunder) is applied to high explosives as an alternative to the word exploding.

There is, too, a difference in usage. The steady thrust from a gunpowder explosion made it very suitable as a propellant in guns or for blasting in mines, and, when there was no alternative, it also served for demolition purposes. But nitroglycerine provided the alternative and its main role was in demolition: in peacetime, for the removal of hard rock in mining and tunnelling and excavating; in wartime, for the filling of mines, bombs and the warheads of shells and torpedoes. It could not be used by itself as a propellant in guns because the rapidity and intensity of the blast would shatter the gun barrel instead of ejecting the cannon-ball or shell.

* * *

The invention of the patent detonator had solved the problem of detonating nitroglycerine in a controlled way, but still no one knew how to prevent it detonating unexpectedly. This issue would never go away, and it reared its ugly head particularly viciously when the small works at Heleneborg was completely wrecked by an explosion on 3 October 1864. Five people were killed, including the Nobels' youngest son, Emil, aged twenty. Immanuel was so deeply affected by the disaster that he suffered a stroke and was bedridden for the last eight years of his life.

For the third time the family faced bankruptcy, but once again they rose to the occasion. Alfred began to pick up the pieces, discarding his brother's advice 'to quit the damned career of an inventor, which merely brings disaster in its train',[14] while Andriette held the family together as she had done thirty years earlier. Even Immanuel was not done for. He wrote three pamphlets on how his mines, which had been used successfully during the Crimean War, could be used to improve Swedish defences, and his fertile mind still produced new ideas, including that of the now commonplace three-ply timber.

The manufacture of nitroglycerine was prohibited within the city of Stockholm, but the explosive had already proved its worth in the building of the new State Railway system, so there was still an active demand for it. Alfred's ingenious response was to restart manufacture on a covered-over barge in the middle of Lake Malaren outside the city limits and to set up his first company, Nitroglycerine Ltd. He was the chief chemist, the works' manager, the secretary and the salesman, yet he managed within a year to build a new factory at a remote spot called Winterviken. The year was 1865, and the factory continued in production for fifty years.

With the Industrial Revolution well under way, and world trade set to double between 1860 and 1875, the time was ripe for expansion and Nobel travelled incessantly to market his exciting new product. In 1865 he founded his first overseas company, building a factory at Krümmel in a remote valley just south of Hamburg in Germany, which could supply European countries. Soon after, he crossed the Atlantic to try to exploit the possibilities in the New World.

Considerable experience in the use of liquid nitroglycerine was being gained in the United States by George M. Mowbray who was using it in the construction of the Hoosac rail tunnel in Massachusetts. This great enterprise had been started in

1848 using hand-drilling and gunpowder, but there had been many setbacks and delays so that even by 1868 only relatively slow progress was being made. At that stage Mowbray began to manufacture nitroglycerine on the site and to blast with that instead of gunpowder. He used it in its liquid form, either loaded into cylindrical tin canisters or cartridges, or poured directly into boreholes. The rate of excavation doubled within two months, enabling the long-drawn-out project to be brought to a successful conclusion in 1876. Mowbray had manufactured, in all, about half a million kilograms of nitroglycerine, and exported some to Canada. That he had achieved this with a fair safety record was mainly due to the fact that he had transported and stored the explosive in the frozen state when it was much less likely to detonate prematurely. In these early days nitroglycerine was also being used in blasting oil-wells to increase the flow of oil.

Many other operators were, however, less successful than Mowbray and liquid nitroglycerine had a reputation for being unsafe. Before he left America, Nobel was chastened to learn that the Krümmel factory in Germany had blown up. This, together with reports of other disasters, led to the passing of the Nitroglycerine Act in 1869, which banned the import of the explosive into Great Britain. It was banned from other countries, too.

CHAPTER 9

Dynamite

The poor safety record of liquid nitroglycerine threatened to involve Nobel in another crisis, so it was particularly timely when in 1867 he made his second great invention. This was a way of taming the liquid nitroglycerine, by absorbing it into a solid to make a paste or dough and, after much experimentation, he decided that kieselguhr was the best material to use as the absorbent. This soft, white, porous substance, made up of the skeletons of minute aquatic plants known as diatoms, occurs widely in many parts of the world, so it was cheap and plentiful. Nobel found that 3 parts of liquid nitroglycerine absorbed into 1 part of kieselguhr gave an easily handled, plastic mixture. It had a lower explosive capacity than the pure liquid because the kieselguhr was inert, but surprisingly it could be detonated more easily. Above all it was safe. Nobel called the explosive 'Nobel's safety blasting powder' or 'Dynamite', from the Greek *dynamis*, meaning power. Any modern advertising agency would have been proud to have chosen the name; for many people it is still synonymous with explosive.

Like Nobel's first invention, the patent detonator, the discovery of dynamite was essentially very simple. In Russia, however, it is claimed that Colonel Petrushevskii, Zinin's collaborator, should be given the credit, and that only the need for secrecy prevented him from applying for worldwide patents. The colonel certainly seems to have been paid 300 roubles and was given a life pension for his work. Had he patented dynamite the courses of both science and history might have been different.

* * *

If the invention of the patent detonator had heralded the era of nitroglycerine in 1864, that of dynamite opened it up in 1867. The demand was overwhelming. In the ten years after the construction of the factories at Winterviken and Krümmel, fourteen others were built in twelve different countries, and the amount of explosives sold rose from 11 to over 5,000 tonnes per year. The build-up of such an empire was not without formidable difficulties. Existing manufacturers of gunpowder were not keen to see their products supplanted so they were ever ready to emphasize the dangers of using the new high explosives, and Nobel had to raise new capital, set up new companies and build new factories. To do all this he had to recruit many assistants and, though he had some good and faithful helpers, some of them turned out to be unscrupulous camp-followers. And he himself had to spend much time on further research and on registering and protecting his patent rights. Dynamite was based on such a simple idea that it was very easy to copy. In one extreme example one devious manufacturer actually renamed a village Nobel – and then described his products as 'genuine Nobel explosives'. Nobel found the existing patent legislation very unsatisfactory, condemning it as made up of 'worm-eaten, misbegotten and stillborn laws' which led to the ' taxation of inventors for the encouragement of parasites'.[1] Nor did he care for lawyers, describing them as 'bloodsuckers who devour fortunes after the delivery of short-sighted interpretations of meaningless court rulings whose obscurity darkens the darkness itself'.[2] And he reckoned that 'lice are a sheer blessing in comparison with journalists, those two-legged plague microbes'.[3]

The Krümmel factory in Germany, demolished by an explosion in 1866, was rebuilt, only to be demolished and rebuilt once again four years later. It grew into Dynamit A/G, which at the start of the twentieth century was the largest explosives firm in Europe. The Krümmel factory employed 2,700 workers during the First World War, and 9,000 in the Second, before it was destroyed in a daylight air raid in April 1945. A bust of Alfred Nobel, damaged but still upright, was found among the debris.

In France, Nobel ran up against a State monopoly in the supply of explosives and his attempts to break into the market were not successful until the Franco-German War of 1870 convinced the French army, rather painfully, that dynamite, used by the Germans but spurned by the French, was highly effective. This led very rapidly to the establishment of Nobel's first factory in France at Paulilles in 1871. A factory was built at Avigliana in Italy in 1873, and Nobel, repaying part of the debt he owed to Sobrero, the discoverer of nitroglycerine, appointed him as an adviser. A bust of Sobrero was erected at the factory in 1879, to perpetuate his name and achievement, and he held his well-paid appointment until his death in 1888.

In England the gunpowder manufacturers would not at first have anything to do with Nobel's discoveries, and the Nitroglycerine Act forbade 'the manufacture, import, sale and transport of nitroglycerine and any substance containing it'. Undaunted, Nobel pressed his case, giving demonstrations to experts as and when he was allowed to do so, and even writing to the Home Secretary who, in the end, was persuaded to license the import and manufacture of 'such high explosive compounds as were proved to be safe'. But Sir Frederick Abel, the chief chemist to the War Department at Woolwich, was trying to do with nitrocellulose what Nobel was doing with nitroglycerine, and he was a powerful adversary with whom Nobel was to cross swords on more than one occasion. So Nobel's attempts to raise money in London failed.

He had better fortune in Scotland and his first joint venture – the British Dynamite Co. Ltd – was founded in Glasgow in 1871. The site chosen for the factory was at Ardeer, a desolate area on the Ayrshire coast, with almost no water supply; Nobel, who spent much of his time there, described the site as having 'barely enough nourishment even for rabbits'. In 1875, after the invention of blasting gelatine, which was also made at Ardeer, the company was reorganized under the name Nobel's Explosives Co. Ltd. Production grew apace and within

An aerial view from the west over the centre of the factory at Ardeer, taken in 1992. (Crown copyright: Royal Commission on the Ancient and Historical Monuments of Scotland)

four years the stock-market value had increased ten-fold. Anything with the name Nobel in it became a firm favourite on the Stock Exchange. By 1890 Ardeer was exporting to many countries outside Europe, particularly to South Africa, Australia and East Asia, and was providing 10 per cent of the world's supply of explosives.

With the reorganization that became necessary immediately after the end of the First World War all manufacturers of explosives in Great Britain had been merged, under the leadership of Nobel's Explosives Ltd, into Explosives Trades Ltd, which changed its name in 1920 to Nobel Industries Ltd. Some 54 companies, with 93 factories, were involved. Similar amalgamation in Germany into a gigantic organization – I.G. Farbenindustrie A/G – had also been taking place, and it was as a result of fierce competition over a wide range of chemical industry from that source in the mid-1920s that Imperial Chemical Industries (ICI) was formed in Great Britain by bringing together Nobel Industries Ltd, Brunner, Mond & Co., British Dyestuffs Co., and the United Alkali Co.

The two main architects of all this rebuilding were Harry Duncan McGowan, later Lord McGowan, and Alfred Moritz Mond, later Lord Melchett. They came from very different backgrounds but both attained positions of great power. McGowan was born in Glasgow in 1874 and went to work as an office boy in Nobel's Explosives when he was fifteen. Twenty-six years later, in 1915, he was elected to the board of directors and became chairman and managing director in 1918. Mond was born in 1868, six years after his father had come to England from Germany. He was educated at Cheltenham College before reading science at Cambridge University and Law at Edinburgh. He was called to the bar in 1894 but only practised for a year before joining the firm of Brunner, Mond of which his father was a co-founder.

Mond was by no means suave, either in manner or speech, and he could be direct to the point of rudeness but, despite his considerable shyness, he was very sincere and kind in handling people. He had great strength of character and was very effective in getting things done because he did not allow himself to be distracted by details. He was a Member of Parliament, on and off, between 1910 and 1928, first for Chester, then for Swansea and, finally, for Carmarthen. He began as a Liberal, and was Minister of Health in 1921 during Lloyd George's coalition government, but his views on the issue of free trade converted him to Conservatism. Throughout his busy life he found time and energy to support the Zionist cause and to nurture his love of art and music. He was created a baronet in 1910 and raised to the peerage in 1928.

Mond was the first chairman and managing director of ICI but after his untimely death in 1930, at the early age of sixty-two, he was succeeded by McGowan. It was the start of a long period of world recession and it needed all McGowan's skills and tenacity to see the company through. He had a ruddy countenance; a fondness for good living; a sound judgement in business, if not always in personal or family affairs; and something of a relish for long, tough negotiation in smoke-filled rooms. His strong will and nerve, coupled with a perpetual optimism, generally enabled him to get his own way, and he arranged a series of agreements with other international chemical manufacturers to ensure a fair share of the declining market for ICI. This necessitated a somewhat dictatorial approach which did not endear him to his fellow directors and, when some rash financial speculation led him close to personal bankruptcy in 1937, the board decided to replace him with the second Lord Melchett. However, before the change could be implemented, Lord Melchett was taken ill and McGowan was asked to continue. He did so until the end of 1951, by which time he had been associated with the chemical industry for sixty-two years. He was knighted in 1918 and created a baron in 1937. He died in 1961. It is probably not too much of an exaggeration to credit him with saving the British chemical industry from collapse in the 1920s and with leading a large part of it through to a bright future.

The explosives part of ICI, called the Nobel division, still operates at Ardeer, but it was sold to ICI, Australia (which is shortly to be renamed Orica) in 1997. In 1955 an ICI historian wrote:

> The Nobel division is proud of its name and of its founder Alfred Nobel. This prince of inventors was a man of rare genius, fecund in ideas, remarkably astute in commercial matters and withal an idealist. From him still springs much of the inspiration for research at Ardeer. If we are to view the flow of explosives research and development as a whole, and not in any particular place, then the real source of the stream is to be found in the attempts made personally by Nobel from 1862 onwards to utilize nitroglycerine.[4]

*   *   *

Nobel was granted a United States patent for 'an improved substitute for gunpowder' on 24 October 1865, but as in England the American powder-makers were very reluctant to admit that there could be any substitute for gunpowder. Dynamite was nevertheless first made, under licence, by the Giant Powder Company in California in 1868. Though it was at first almost ten times more expensive than gunpowder it was soon found to be much more effective in most mining operations, and all the gunpowder manufacturers began to realize that their days were numbered unless they also began to manufacture the new high explosive.

The California Powder Works entered the field in 1874, manufacturing a type of dynamite which was sold under the name Hercules powder. At that stage, du Pont de Nemours & Company intervened, Lammot du Pont having at last persuaded his Uncle Henry that he must move away from gunpowder if the company was to flourish. They began by investing, somewhat surreptitiously, in the California Powder Works and by building up a very fruitful liaison with them through Lammot. It was in this way that he learned enough about dynamite to be chosen to establish and run a completely new organization, the Repauno Chemical Company, which was set up jointly by du Ponts with Hazard and Laflin & Rand in 1880. Henry du Pont, who was still sceptical and suspicious of the whole idea, stipulated that any new factory must be remote from the Wilmington site. The company's buildings were, therefore, erected, some 50 km away, at Gibbstown in New Jersey on the opposite side of the River Delaware. As Lammot

Lammot du Pont. (The Hagley Museum and Library)

du Pont cut the first sod he said: 'This will one day be the biggest dynamite plant in America.'[5] It was not long before it was the biggest in the world.

Their first dynamite product was called Atlas powder and the success of the venture was demonstrated by a six-fold rise in production from 228 tonnes in 1880 to 1,372 tonnes in 1881. By then the manufacturers of Hercules powder had been bought up, and other types of dynamite were marketed under such names as Rend Rock, Vigorite, Nitroleum and Ajax Powder. The 'Dynamite' name was never used to avoid possible patent infringements.

In the boom and take-over atmosphere that existed at the time there were many success stories but there were also two important casualties. Nobel, who had started it all off and who had the original patent rights in the United States, was confronted with so much sharp practice, so many patent infringements and so much litigation that he grew weary of it all. He so disliked the American way of life that in 1885 he sold all his interests for a mere $20,000, vowing never to visit the United States again. Nor did he.

Lammot du Pont's fate was even worse, for he was killed by an explosion in his own plant on 18 March 1884. A load of 1,000 kg of nitroglycerine had been left overnight in a lead-lined vat, but in the morning it was found to be overheating and fuming. All might have been well if the workmen on the spot had run the fuming liquid into the adjacent tank of cold water placed there for just such an emergency but instead they sent for help. When Lammot and an associate called

Hill arrived it was too late. Killed instantly as the vat blew up, they were buried under a great pile of debris. Lammot was buried in the family cemetery on the site of the original gunpowder mill at Wilmington and the grave was marked by a broken stone column, the customary headstone for one who had died tragically. Once again, the du Pont family had to 'close up and march on'. Lammot's eldest son, Pierre Samuel, aged only fourteen, informed his mother that 'I am the man of the family now.'[6]

* * *

The discovery of gold in California around 1850 boosted the demand for explosives and encouraged the establishment of dynamite manufacture in that region. Likewise, the discovery of gold and diamonds in South Africa, some forty years later, triggered off similar developments there. In 1895 the Nobel organization in Europe acquired a controlling interest in a South African explosives company, the Zuid-Afrikaansche Fabrieken voor Ontplofbare Stoffen, and began to build the largest dynamite plant in the world at Modderfontein ('Muddy Springs') in the Transvaal. Opened by President Kruger in 1896, it rapidly established a powerful monopoly, which it maintained until the outbreak of the Boer War in October 1899 changed everything.

Most of the mines were closed down for the duration of the war and the works at Modderfontein were quickly adapted to provide explosives and munitions to the Boers. Throughout 1899 the Boers had considerable success, besieging Ladysmith, Kimberley and Mafeking but, with the appointment of Lord Roberts as the British Commander-in-Chief and a large increase in the number of his troops, towards the end of that year, the British forces slowly began to gain the upper hand. They captured Modderfontein in June 1900 and the town became the headquarters of the newly formed British peace-keeping force – the South African Constabulary – under the command of Major-General Baden-Powell. He was already a popular hero following his successful command as the colonel of the 20 officers and 680 men who withstood the 217-day siege of Mafeking, which was not relieved until May 1900. Despite other British victories and the capture of such important centres as Bloemfontein, Johannesburg and Pretoria in 1900, the Boers carried on a guerrilla war until a final peace treaty was signed at Vereeniging in May 1902.

After the war, the assets of the old Modderfontein organization were transferred to a new company, the British South African Explosives Company, and an experienced Scotsman, William Cullen, who had worked at Ardeer and at Waltham Abbey before becoming a freelance consultant in London, was sent to re-establish the old operation. He worked wonders in the chaotic situation which prevailed as the mines opened up after three years of conflict, but the old days were never to be resurrected. The mining companies had always resented Nobel's monopoly position at Modderfontein and, just before the start of the war the largest company in South Africa, De Beers Consolidated Mines, whose chairman was Cecil Rhodes, had decided to start manufacturing dynamite themselves. To this end, they persuaded Colonel W.R. Quinan, a former United States Army officer who had become a manager for the Californian Powder Works, to move to South Africa. He was reluctant to leave a job which he had held for fifteen years and in which he was very happy, but De Beers met all his personal demands and more or less promised him a blank cheque to build a new factory.

He arrived in South Africa in 1899 and chose a site at Somerset West, near Cape Town. With the help of his son and some experienced workers he had brought with him, he planned to have the factory built within two years but he had not taken into account the disruption caused by the war and he was also impeded by considerable political and social opposition to his plans. It was, therefore, the middle of 1903 before the De Beers Explosives Works dispatched its first consignment of dynamite.

South Africa now had the two largest dynamite plants in the world and there was some stern competition between the two great organizations that backed them. The cosy monopoly position had certainly been broken. And the competition became even fiercer when Kynoch's, Nobel's biggest competitor in the United Kingdom, decided that they would also enter the fray by building a works at Umbogintwini, near Durban. By 1909, then, there were three companies employing, between them, over 3,000 people, and making explosives was the largest manufacturing industry in South Africa. The competition almost halved the price of dynamite but, luckily for all three firms, the demand for explosives had almost trebled in a decade so that they were all able to make some profit.

There was nevertheless excess capacity and vague suggestions of amalgamation began to be aired, but it was not until after the First World War, in which the South African explosives factories played only a very small part, that any remedial steps were taken. Then the rationalization occurring all over the world began to affect the South African interests. Following the absorption of Kynoch's into Nobel Industries Ltd, in the United Kingdom, the Umbogintwini plant stopped making explosives in 1921 and was converted to the manufacture of fertilizers and other agricultural products. Some acrimonious negotiations followed between Nobel Industries and De Beers, with each side distrusting the other, until a new jointly owned company, African Explosives and Industries Ltd, was formed in 1924. By then the Somerest West factory, on the advice of Kenneth Quinan, who had become manager when his uncle, W.R. Quinan, died in 1910, was also turning its attention to manufacturing fertilizers.

This move away from explosives into more general chemical fields went on apace and was reflected by a change in the name of the company, in 1944, to African Explosives and Chemical Industries Ltd. By then it was the main manufacturer of South Africa's chemicals but it continued to make explosives and it greatly helped the British war effort between 1939 and 1946 by providing about sixty experienced workers to assist in training and management in the new explosives works which were built in the United Kingdom.

* * *

Nobel's original dynamite mixture came to be known as kieselguhr dynamite, guhr dynamite or dynamite No. 1, but the word dynamite soon became a generic term for a family of well over a hundred mixtures all, it was hoped, better in some way than the original. All dynamites contain nitroglycerine; it is just a matter of changing the base or the dope with which it is blended. Dynamite No. 1 is classified as a dynamite with an inactive base because the kieselguhr plays no real part in the explosion except to dilute it. This type of dynamite, possibly using different absorbents, was widely used in Europe for a time but not in the United States.

It was not long, however, before it was realized that more effective mixtures could be made by replacing the inert base, at least partially, by a substance (or a mixture of substances) which would itself burn and/or explode. Nobel had patented such a mixture in 1869. It contained nitroglycerine, resin or charcoal, barium nitrate and sulphur. There were endless possibilities, with mixtures containing potassium nitrate being the most popular, at least to start with. Dynamite No. 2, containing 72 parts of saltpetre, 10 of charcoal, 18 of nitroglycerine and a little paraffin, was milder than No. 1 and very suitable for use in most quarries and coal-mines; Dynamite No. 3, a mixture of equal parts of No. 1 and a mixture of saltpetre and wood-meal, was intermediate in power between No. 1 and No. 2; Atlas powder, used in the construction of the Panama Canal, contained nitroglycerine, wood pulp, saltpetre and magnesium oxide or chalk; Rend Rock had nitroglycerine with saltpetre, wood-meal and pitch; Hercules powder contained nitroglycerine, saltpetre, magnesium carbonate, potassium chlorate and sugar.

In America sodium nitrate and ammonium nitrate were widely used instead of saltpetre. Judson powder contained nitroglycerine, sodium nitrate, sulphur and bituminous coal. A mixture of 60 per cent nitroglycerine with sodium nitrate and wood-meal was known as a 60 per cent straight dynamite, and the percentage could be lowered down to 5 to give a range of mixtures with different powers. When some of the nitroglycerine was replaced by ammonium nitrate the products were known as ammonia or extra dynamites.

Lithofracteur, with a composition very similar to the mixture patented by Nobel in 1869, was a modifed dynamite with an interesting history. It was invented by a German, Friedrich Krebs, but when he tried to introduce its manufacture into Great Britain in the early 1870s he was sued by Nobel for patent infringement. After four years of litigation, ending up in the House of Lords, the firm of Krebs Bros & Co. was forced to close down in Britain and to remove all their existing stocks. But, like Nobel, Krebs did not give up easily. He turned to Australia. The population of that country, at the time, was only about two million, but the increasing activity, particularly in mining the gold which had been discovered around 1850, made it a good market for explosives. The newly established Lithofracteur Company (Krebs Patent) Ltd was allowed to operate unchallenged and set up its factory outside Melbourne on a site known then as Kororiot Creek. This is nowadays Deer Park, an important maufacturing complex of ICI Australia Ltd, which will shortly change its name to Orica. This is just one more example of how explosive manufacture provided the early base from which much wider chemical industries developed.

*   *   *

An 1880s advertisement for the Australian Lithofracteur Company. (Deer Park Historic Archive, ICI Australia Limited)

Many of the modified forms of the original dynamite were very effective but there was always likely to be some separation of the liquid nitroglycerine from the base, particularly in the presence of moisture, or under pressure, or on prolonged storage. This turned Nobel's thoughts towards trying to find something that would dissolve in nitroglycerine instead of just absorbing it or mixing with it. Following the well-known general principle that 'like dissolves like', Nobel at first tried to blend nitroglycerine with nitrocellulose, which had been discovered in 1846 and which was known to be an explosive, but without success as they would not mix satisfactorily. But initial failure never deterred him. As he once wrote: 'I always return to anything of which I have the feeling that I shall succeed with it in the end.'[7] His genius was both inspiration and perspiration. So it was that he cracked the problem some eight years later.

Legend has it that he cut his finger one day and applied a solution of collodion – a partially nitrated cellulose – to form a protective skin over it. As the aching finger kept him awake during the night he had plenty of time to ponder and it dawned on him that this form of nitrated cellulose might be better than the more fully nitrated form that he had tried eight years previously. He rushed down to his laboratory at 4 o'clock in the morning and, within a few hours he had made his third great invention.

Incorporation of 7–8 per cent of collodion into nitroglycerine forms a stiff, jelly-like mass which Nobel called blasting gelatine. At first he used a solvent, such as acetone, to facilitate the blending of the two materials but he later found that solution could be achieved simply by adding the collodion to warm nitroglycerine. Blasting gelatine was patented in 1875 and it proved to be a very powerful explosive, so much so that it was used as a standard against which to measure the strength of other explosives. It was cheap to manufacture; it was more powerful than dynamite because both the nitroglycerine and the collodion which it contained were explosive; it was unparalleled for blasting hard rock; it was insensitive to shock; it could be packed very well into boreholes because of its jelly-like nature; and it was so resistant to moisture that it could be used under water even after long immersion. It was, however, too powerful for all but rather special uses.

But there was a closely related alternative, because lowering the collodion content to 2.5 per cent produced a viscous mass that could be absorbed into a mixture of nitrates and wood-meal to make gelatine dynamite, which could also be used under water. One of the most common varieties, using potassium nitrate as the nitrate, was called gelignite. NS gelignite contains sodium nitrate instead of the potassium salt, and ammon gelignite contains ammonium nitrate. A selection of trade names – Duxite, Super-Cliffite, Express Dynamite, Essex Powder, Geloxite, Rippite, Samsonite and Saxonite – gives some idea of the variety of product that has been marketed.

*　　*　　*

The availability of these improved explosives – more powerful than gunpowder and safer than pure nitroglycerine – gave a great boost to mining and civil engineering. They were at first very expensive, partly because Nobel's companies had a monopoly in their manufacture. In England dynamite cost about six times as much as gunpowder. The new explosives were nevertheless rapidly adopted, and even more so after the price fell in 1881, when the original patent rights ran out. Nobel's companies had claimed that the rights should be renewed on the grounds that they had not had long enough to make adequate profits, but this claim was successfully contested, mainly by Cornish mining interests. One mining journal reported: 'if everyone connected with dynamite has not made his fortune, it has not been for want of knowing how to charge'.[8] The explosive had in fact been twice as expensive in England as in Germany.

By 1872 gunpowder and the unsafe liquid nitroglycerine had both been replaced by dynamite in all underground mining in Cornish tin and copper mines, and it was the introduction of this new explosive, together with power drilling, that at least postponed the demise of the industry. There was even a boom in 1881 but the sad decline in the percentage of the world market taken by Cornish tin from 37 per cent in 1860 to 1.5 per cent in 1920, and the increasing emigration of Cornish miners, could not be halted. The competition, first from Malaysia and Indonesia, then from Australia, and finally from Bolivia was too much.

Nor was dynamite without its faults. It was tried in the early days of excavating

A variety of Nobel cartridges of blasting explosives in 1907. (Crown copyright: Royal Commission on the Ancient and Historical Monuments of Scotland)

the Severn tunnel but it did not perform well in the cold and the contractor wrote that 'it produced such deleterious and even dangerous fumes that it was abandoned altogether'. The acrid fumes it produced in poorly ventilated mines in Cornwall were so obnoxious that it was 'not unusual for men to come six days to work and vomit their breakfast every morning . . . and not be able to eat their dinner'.[9] It was also necessary to treat the new explosive with even greater respect than gunpowder but the Cornish miner, inured by constant accidents, was almost fatalistically independent and happy-go-lucky. One miner was killed when carrying a stick of dynamite and a candle in the same hand; another blew off the side of his house when he tried to use the explosive for cleaning his chimney; and, when 25 kg of dynamite was used at the West Bassett mine to bring down 1,500 tonnes of ground, which would have taken several years by hand labour, the rush of air was such that men were swept off their feet and injured at the end of the level where they had thought themselves safe.

In civil engineering the new explosives opened up a completely new era, with many remarkable projects being undertaken. The St Gotthard pass route, with 324 bridges and 80 tunnels, was completed using dynamite and blasting gelatine between 1872 and 1882. Other great enterprises around the same time included the Suez Canal (1859–69); the beginning of the London Underground system with a 6.5 km underground railway line between Bishop's Bridge Road in Paddington and Farringdon Street on the city boundary (1860–3); a 15 km tunnel through granite from Goschen to Airolo in Switzerland (1872–80); the opening up of Hellgate in New York's East River (1876 and 1885); the Corinth Canal (1881–93); the clearing of the River Danube at the Iron Gates (1890–6); the

The Corinth Canal in Greece. The canal, which links the Ionian to the Aegean Sea, took ten years to build and was opened on 6 August 1893. It is about 6 km long and 23 m wide. The vertical rock walls rise 90 m above the water level. (The National Tourist Organization of Greece)

Simplon Railway tunnel (1898–1906); and the Panama Canal (1904–14). The possibilities were romanticized by a contemporary lyricist:

> Do you wish to make the mountains bare their head
> And lay their new-cut forests at your feet?
> Do you want to turn a river in its bed,
> Or plant a barren wilderness with wheat?
> Shall we pipe aloft and bring you water down
> From the never-failing cisterns of the snows,
> To work the mills and tramways in your town
> And irrigate your orchards as it flows?
> It is easy! Give us dynamite and drills!
> Watch iron-shouldered rocks lie down and quake
> As the thirsty desert-level floods and fills,
> And the valley we have damned becomes a lake.[10]

* * *

Alfred Nobel was a thin man, below medium height, with deep-set eyes and a prominent nose, on which in later years he used to hang a pince-nez. Even growing a beard, when he was about thirty years old, did not conceal his rather pale face and distinctly melancholy, staring look, which was matched by his sad, tense voice. He was never robust nor self-confident and was often in poor health, which makes his achievements all the more remarkable. He was not only a great inventor, he was also a great worker, a great salesman, and a great organizer. By the time of his death in 1896, aged sixty-three, there were ninety-three factories throughout the world, all operating under his banner, producing 67,500 tonnes of explosives annually. The annual production had been 11 tonnes thirty years earlier. And when he died, his estate was valued at 33,000,000 Swedish kronor (about £2,000,000). What a man! But what sort of a man? The simple adjectives must be complex, cosmopolitan, unassuming and, above all, lonely.

The complexity arises because he was a man of marked contrasts. He was physically frail yet he had an enormous capacity for hard, prolonged work, for making decisions and for overcoming obstacles and defeats. Although he spent most of his life dealing with explosives and armaments he was a man of peace. 'For my part,' he wrote, 'I wish all guns with their belongings and everything could be sent to hell, which is the proper place for their exhibition and use.'[11] In many of his business dealings he had to be tough, and he was not afraid to be so, yet he was very sensitive and easily hurt. He was, for the most part, calm and detached but he had a bad temper. He wrote, 'When the Nobel blood surges up, there is no lack of explosiveness. I get so angry that the sparks fly – but it lasts for only half an hour.'[12] He was a scientist yet he had a deep literary interest. Despite his hectic life he wrote three unpublished novels and a play, but his main interest was in poetry which he both read and wrote extensively. Shelley was a particular favourite. And his letters reveal a telling choice of word and phrase, bringing his personality to light, even though he frequently seemed to write with tongue in cheek in order to shock. He liked the use of aphorisms: 'Contentment is the only real wealth', 'Worry is the stomach's worst poison', and 'Hope is nature's veil for hiding truth's nakedness' show what he was capable of.[13]

His enforced lifestyle ensured that he was cosmopolitan. After his early days in Sweden and Russia, he lived in Germany, France and Italy before returning to Sweden towards the end of his life. He travelled all over Europe, though he went only once to the United States, and perforce he came to speak five languages – Swedish, Russian, French, German and English. In his business activities he had to take a worldwide view, but he always remained a patriotic Swede, never forgetting his early upbringing, and he visited his mother every year on her birthday. Nor were some other countries entirely to his liking. Of England he wrote, 'Conservatism is too flourishing for counsel to accept anything which has no antediluvian sanction', and of France, 'All Frenchmen are under the blissful impression that the brain is a *French* organ.'[14]

Though he was such a famous and important person, he always preferred to take a back seat outside the business world, and his modesty was in no way false. Asked to launch a new ship that was to be named after him, he replied, 'Since she is both elegant and shapely it would be a bad omen to name her after an old wreck.'[15] As a contribution to a family biography being compiled by his brother he offered the following: 'Alfred Nobel – pitiable half-creature, should have been stifled by humane doctor when he made his entry yelling into life',[16] and rated important events in his life as zero. It was, however, probably his loneliness that most affected him and made him so sad and melancholy. He revealed his greatest fault as 'lacks family, cheerful spirits, and strong stomach',[17] and he replied to someone who had referred to his many friends, 'Where are they? On the muddy bottom of evaporated illusions or close to the clatter of piled-up coins?'[18]

The adoration and respect that Nobel had for his remarkable mother made him feel the lack of his own family very deeply, and it was probably the lack of any permanent female companionship that contributed most to his loneliness. Yet though he regarded himself as ugly and unattractive, he did have some interesting, if temporary, female relationships. In his first overseas trip from Russia, when he was still very young, he fell deeply in love with a girl in Paris, but their mutual happiness was ended by her tragic death during the short time he was there. It was in Paris again, when he was forty-three, that he advertised for a secretary and eventually employed Countess Bertha Kinsky. She was clever and beautiful, and Nobel was certainly interested in extending the relationship beyond the office, but, within the year, she had left to marry someone else.

He hoped he had at last found a partner in Sofie, a vivacious twenty-year-old whom he met in a flower-shop while on a business trip in Austria, but there was to be no permanent joy even though the liaison lasted for eighteen years. He was more than twice her age, clever, cultivated, mature and somewhat sedate. She was quite uneducated and had run away from an unhappy home in search of bright lights and fortune. Nobel brought her to Paris and set her up in an apartment close to his bachelor flat. He lavished all his attention upon her, felt sorry for her, and tried with all his might to satisfy her every whim, but she was too giddy and indolent and all his efforts to refine and restrain her were to no avail. She used his name, she took more and more of his money, she constantly tried his patience and she subjected him to much ridicule. In his late fifties he wrote to her: 'My whole life turns to bitter gall when I am forced to act the nanny to a grown-up child and be the butt of all my acquaintances.'[19] Sofie's response was to live it up even more wildly and the final blow fell when in 1891 she had a child by a young Hungarian officer. That might have been the end of it but, though Nobel granted her a generous annuity, she persisted in trying to prise more money out of him, even after his death. Her deplorable behaviour, coupled with his brother Ludvig's death in 1888 and his mother's in 1889, added greatly to Nobel's isolation in the last few years of his life.

* * *

Countess Kinsky, Nobel's former secretary, blossomed as an author under her married name of Bertha von Suttner, writing passionately on behalf of the peace movements of the day and becoming quite famous. Although she only saw Nobel two or three times after she had left his employ she corresponded regularly with him to try to attract his financial and moral support for her cause. Nobel was not averse to such approaches for he had always been more interested in peace than war. He wrote in 1886 that he had 'a more and more earnest wish to see a rose-red

The demolition of a block of flats in Glasgow. The time lapse between photographs is approximately 6 seconds. (Robinson & Birdsell Ltd, Leeds)

peace sprout in this explosive world',[20] and his thoughts turned increasingly to peace as he grew older. So much so that, aged sixty, he wrote: 'I should like to leave part of my fortune to a fund for the creation of prizes to be awarded every five years (let us say six times, for if within thirty years one has not succeeded in reforming society such as it is today, we shall inevitably relapse into barbarism) to the man or woman who has contributed in the most effective way to the realization of peace in Europe.'[21]

Nobel had made his first will in 1889 during a period of ill-health; he made a second one in 1893; and his final one a year before he died. It was dated 27 November 1895 and was drafted in Swedish in his own hand without any legal assistance. It was clearly intended, in good faith, to launch the Nobel prizes for physics, chemistry, physiology or medicine, literature and peace, but it must serve as a warning to anyone who is tempted to draft his or her own will, to leave so much money, or to award international prizes. It took more than five years of legal wrangles before Nobel's wishes could be implemented. He would not have liked it. The will had not been well drafted; the precise domicile of such a well-travelled man was not legally clear; the money was tied up in at least nine different countries, all with their own laws; the institutions named as the awarders of the prizes had not been forewarned of their role and were unhappy and uncertain about what they were expected to do; there was no clearly defined machinery for administering the finance or the prizes; and many jingoistic Swedes were angry that the prizes had been made international and that the Norwegian Storting had been chosen to award the prize for peace. On a more personal note, some of Nobel's relatives tried to get their share of the estate increased, and the executors even had to buy Sofie off when she threatened to sell over 200 of Alfred's love letters if she didn't get more money. It was 29 June 1900 before the statutes governing the Nobel Foundation were finally promulgated, since when the Nobel prizes have been a glittering diamond on the sometimes murky historical scene, but not without continuing controversy over some of the awards.

*   *   *

Nobel left Paris in 1891 after living there for eighteen years, and went to San Remo in Italy. His new villa was called Mio Nido ('My Nest') but when a friend pointed out that a nest should be occupied by two mates the name was changed to Villa Nobel. He was by no means inactive, for that was not his nature, but he was tiring and he was saddened, once more, by the death of his brother Robert in 1896. Alfred had suffered some heart trouble earlier in his life, and he now began to complain of cramp and headaches, having what he referred to as 'repeated visits from the spirits of Niflheim'.[22] He wrote to a friend: 'Life is short and mine will be particularly so.'[23] What irony that French specialists should diagnose angina pectoris and that he should be treated with nitroglycerine, the oil which he had tamed and harnessed but which could do nothing for him in his hour of need. He died on 10 December 1896. He was only sixty-three.

CHAPTER 10

Guncotton

Cellulose, which is the main constituent of all plant cells, is the most abundant naturally occurring organic chemical. It is found in particularly large quantities in fibrous materials such as flax, hemp, bamboo, grass and coconut fibre, but the main commercial source for many years was cotton, which was replaced to some extent by wood and straw when it became too expensive. Although cellulose is so commonplace, it has a complex arrangement of atoms in its molecule; this was elucidated in 1937 by Sir William Haworth, Professor of Chemistry at Birmingham University, whose pioneering research won him the Nobel Prize for Chemistry.

A year before Sobrero had first made nitroglycerine (NG), a German chemist, Schönbein, had treated cellulose with a mixture of hot concentrated nitric and sulphuric acids to make nitrocellulose (NC). He used cotton wool as his source of cellulose, and as he found that the nitrocellulose he made was explosive he called it guncotton. It is, however, unfair to give him all the credit for the discovery because a number of other chemists, notably Rudolf Böttger (often regarded as the inventor of matches) and Jacob Berzelius, all have good claims. The knowledge of chemistry and its dissemination at the relevant times were so uncertain that it was not at all clear who was doing what.

*   *   *

Christian Friedrich Schönbein, born on 18 October 1799 at Metzingen, was the oldest of eight children. His father was a dyer but when his business declined he gave it up and entered the postal service. He was not very rich and could not provide much schooling for all his children so, at the age of thirteen, Christian started an apprenticeship in a chemical firm some hours' journey away from his home. He was young and very homesick but he was clever and he immediately began to take advantage of the opportunities to learn not only chemistry and physics but also Latin, French, English, philosophy and mathematics. By the age of twenty-one he had made such good progress that he moved to another chemical factory near Erlangen so that he could be close to the university there, and he so impressed his new employer that he was appointed as personal tutor to his children. This provided him with enough money to enter the university, in May 1821, with the intention of training to be a teacher.

He began his chosen career by teaching in an educational community

C.F. Schönbein. (Anne Ronan Picture Library)

organized by Friedrich Froebel, the founder of the now famous kindergarten movement, and he taught for a while at a boys' school in Epsom, England, and in Paris. But his life changed dramatically in 1829 when he was offered a temporary appointment to stand in for an ailing professor of chemistry and physics at the University of Basle, and this became a permanent post when the professor's illness did not respond to treatment. It was at the University of Basle that Schönbein discovered guncotton, and he remained as professor there until he died on 19 August 1868.

He was described by one friend as 'small in stature, very fat, mobile as a ball, and full of bubbling humour',[1] and by another as 'unconventional, with an unspoiled, original freshness'.[2] His humble origins were reflected in a somewhat naive simplicity, particularly in business affairs, and in his candour, shyness and complete lack of sophistication. Above all, he was a sociable, homely character, still feeling, at the age of sixty-five, 'the need of occasionally breathing native air, of eating sauerkraut, black sausages and dumplings, and of drinking Neckar wine'.[3] Yet he could be stubbornly independent and had a mind of his own. As a chemist he was something of an agnostic, pooh-poohing the ideas of atomic theory and the vogue for quantitative measurement and organic chemistry which were all fashionable in his day. He preferred to pursue his own interests, which were mainly centred on oxygen, and he did this with a keen insight and much originality and ingenuity. He produced more than 350 papers in his lifetime, including one announcing the discovery of ozone in 1839.

Legend has it that Schönbein occasionally carried out experiments in the kitchen at home, even though such activities had been expressly forbidden by his wife. One day, so the story goes, a flask in which he was heating a mixture of concentrated nitric and sulphuric acids broke, spilling the acid mixture all over the floor. The first thing he could find to mop up the mess was his wife's cotton apron, but when he came to dry it in front of the stove it caught fire and exploded. His wife's remarks are not recorded. However, this is an unlikely tale because there is good evidence that Schönbein was led to his experiments with cellulose by a train of thought arising out of his work on oxygen.

The guncotton that he made looked very much like the cotton wool he had started with, but it felt harsher. When ignited in the open it burnt with a flash but in a confined space, it exploded. It was, then, very much like gunpowder, but it was noticeably more powerful. He sent samples to Faraday, Herschel and Grove in England describing the material as a dangerous rival to gunpowder, and he wrote to Dumas:

> You know, perhaps, that I have discovered a very simple method of transforming cotton into a material possessing all the necessary properties as propellant. In addition to the superior explosive force of this curious substance, it is in every respect superior to the best powder. Experiments which I have made in mines and quarries and with cannons and mortars have shown that 1 pound [0.45 kg] of this substance produces equal effects to that from 2 to 4 pounds [0.91 to 1.81 kg] of ordinary black powder. It should be added that cotton so treated does not leave any residue when exploded, and produces no smoke. The manufacture is not attended with the least danger and does not require costly installations. In view of these properties we cannot doubt that this explosive cotton should rapidly find a place in the pyrotechnic arts and especially in war vessels.[4]

Some of the hyperbole was the myopic enthusiasm of an inventor, but he clearly recognized the importance of his discovery, so much so that he kept the secrets of how to make guncotton to himself, though he did make a general, vague announcement to a scientific meeting in Basle on 27 May 1846. And with none of Sobrero's qualms, he immediately set about trying to exploit the commercial and military potential of the new material. He had never been at all materialistic, writing, while still a young teacher, that 'money is not one of my first desires'[5] and commending any endeavour 'to fill the short span of our lives with deeds of lasting value'[6] as a noble ambition. Yet at this important moment of his life he must have had some hopes of great wealth.

*   *   *

In England, Grove had already introduced guncotton to the British Association at their meeting in Southampton in 1846, and Schönbein arrived in August to add to the salesforce. He gave demonstrations at Woolwich, and was received by Queen Victoria and the Prince Consort, to whom he presented the first brace of partridge to be shot using guncotton. So much interest was aroused among gunpowder manufacturers, mining companies and military establishments that the government granted £1,500 to pay for further demonstrations. At that time Cornwall was one of the world's major mining areas, and Schönbein, accompanied by a senior partner from the international mining firm of John Taylor & Sons, went to the Penryn granite quarry to show off the blasting powers of guncotton. The local hard-headed miners were highly sceptical that what looked just like cotton wool could blow anything up. One even offered to sit on top of the borehole when it was fired if rewarded by a pint of beer. But scepticism soon turned to amazement when they saw what the guncotton could do, and they quickly realized that it was indeed far more effective and much cleaner than gunpowder.

Schönbein agreed to pass on the details of how to make the explosive to the Taylor firm and these were revealed for the first time in a patent[7] taken out by John Taylor in October 1846 to cover,

> Improvements in the manufacturing of Explosive compounds communicated to me from a certain foreigner residing abroad. The invention consists of the manufacture of an explosive compound applicable to mining purposes, the throwing of projectiles, or otherwise, as a substitute for gunpowder, by treating or combining matters of vegetable origin with acids. The vegetable matter which is found best suited for the purpose of the invention is cotton. . . .

The cotton, after cleaning and drying, had to be immersed in a mixture of 1 part of concentrated nitric acid and 3 parts of concentrated sulphuric acid, at 15°C, for 1 hour. The product was then washed thoroughly until free from acid, pressed as dry as possible and then spread out and warmed to 65°C for final drying.

Taylor signed an agreement with John Hall & Sons, well-established manufacturers of gunpowder at Faversham, giving them the sole rights to operate the process for three years. In return they were to pay one-third of the net profit, with a minimum down-payment of £1,000 and the same amount annually. Halls built the first ever factory for making guncotton at Faversham and advertised their new product as being six times more powerful than gunpowder.

It was, alas, soon to show its power, for on 14 July 1847 two buildings in the new factory were completely demolished and twenty-one people were killed. An eyewitness account was published in *The Times*:

> The roofs of all the buildings within about a quarter of a mile [0.4 km] of the explosion are completely stripped of their tiles, and the walls are much shaken. Even in the town of Faversham, fully a mile [1.6 km] distant from the scene of the disaster, windows were broken, and the houses otherwise damaged in some instances. On the opposite side of the stream which forms the northern boundary of the Marsh Works is a field of wheat of some extent. The explosion has completely blasted this over a space of about two acres [8,100 sq. m], and

The aftermath of the explosion at Hall's Marsh Works in Faversham on 24 July 1847 (From The Illustrated London News, *24 July 1847)*

the ears, drooping and discoloured, present a scene of desolation in perfect character with the adjoining ruins. The willow-trees which skirt the bank of the stream referred to, and, indeed, all the trees within about fifty yards [45 m] of the buildings Nos 3 and 4, are torn up by the roots, and scattered about in all directions.[8]

John Hall & Sons wrote to Schönbein in August:

The circumstances attending the late awful explosion of our guncotton establishment and the awful sacrifice of life connected with the destruction of so much property, have so overwhelmed us with trouble, and difficulty, that we have hardly been able to settle our minds, so as to be able to make any detailed communication to you on the subject . . . Eighteen persons were killed by that explosion, ten only could be recognised, the remainder were literally blown to atoms and scattered with the materials in every possible direction. One other person who inhaled the fumes of the acid, and who acted incautiously in not attending to medical advice, also died on the evening of the explosion. Of the survivors, fourteen in number, who suffered dreadfully by broken limbs, contusions, and being burnt by the acids, one has since died and we fear one or two more will hardly recover. Some are maimed, and we are obliged by principles of sympathy to maintain them, and furnish medical advice and assistance.[9]

There were also some awkward financial details, for the letter goes on:

This calamity, attending with all its trying and appalling circumstances, has exceedingly embarrassed our position and has placed us in the situation of submitting to you the first moment we could get in the accounts, a balance-sheet and the divisible third, which we place to your debit, in this painful matter . . . You may rest assured of the deep anxiety we have in working out this affair in the best way we can to recover some portion of the severe loss we have incurred, and that it remains with you to facilitate and promote your views in every possible way in your power, for we must confess to you that we should never have entered into any agreement or had anything to do with the matter had we not relied on your express declaration that the guncotton could be made for *tenpence sterling* per pound, a price assumed on fallacious data and wholly at variance with experience and facts. [10]

The cause of the disaster was never ascertained, and Schönbein had neither the money nor the inclination to start a legal battle but it had not been a very propitious entry for him into the commercial world. He replied to Hall's letter in October, suggesting that the guncotton had been overheated in the drying process and expressing his condolence:

Though I have already expressed to you my deep sorrow regarding the disastrous explosion, I cannot help returning once more to that melancholy subject and assuring you that no event in my life has ever produced such a deep and painful impression upon my mind as that to which I refer. You know that my humour is rather of a cheerful turn, but ever since I got that afflictive piece of intelligence my spirits have been saddened, and you may easily imagine how deeply I feel for Messrs Hall and in particular for him who had to witness the catastrophe and nearly fell a victim to it. It must have been the most trying moment in which a man can be placed in his life; and the killed and the wounded! I will stop here and confine myself to expressing my heartfelt wishes that it may please kind Providence to spare Messrs Hall a renewal of such hard trials.[11]

While this rather quaint correspondence was going on, there were other guncotton explosions at Vincennes and Le Bouchet, and these events put an end to the manufacture of the material in both England and France for about sixteen years.

Meanwhile, Schönbein had been active in other parts of Europe and had offered his process to the Deutsche Bund for 100,000 thalers, but a committee appointed to consider the matter turned the offer down after six years of deliberation. Thereupon, in 1852 Baron von Lenk, who had been the Austrian representative on the committee, and its secretary, persuaded the Austrian government to buy it for 30,000 guilder. He adapted Schönbein's method of manufacture by using skeins of cotton yarn and by trying to ensure that all excess acid was removed before the guncotton was heated in the final drying. He did this by washing it in water for three weeks and then boiling it with a solution of potassium carbonate. He carried out this process at a factory in Hirtenberg and for a few years it looked as though he had hit upon a safe way of making guncotton that could be used for blasting, even if he could not produce it in a form that was satisfactory as a propellant. But it was another false dawn and on 30 July 1862, Hirtenberg joined the list of factories destroyed by internal explosions. When there was a further explosion at another factory near Vienna in 1865, the manufacture of guncotton was stopped in Austria.

This enabled the persistent von Lenk to give more attention to his activities in France, America and England. He took out patents in England, in 1862 and 1863, in the name of Révy, and became an adviser to the firm of Messrs Thomas Prentice & Company when they built a new factory – the Great Eastern Chemical Works – at Stowmarket in Sussex. Production began on 26 June 1864, the guncotton being marketed in rope-like lengths spun from fibres of nitrocellulose. They could be ignited by safety fuse, and they began to be used very successfully in Cornish mines alongside gunpowder, often being stored under water until required for use. It was quite common for the more expensive guncotton to be used in the bottom of a borehole which was then filled up with gunpowder.

So there was progress at last, but it was eighteen years since Schönbein, now aged sixty-six, had set foot in England so hopefully, and a lot of time had been lost. He died three years later, quite unexpectedly, a somewhat disillusioned man. 'In one sense', he had written to his wife, 'it is a misfortune to make a discovery which has a practical importance; it disturbs your peace of mind to the utmost.'[12]

By then, guncotton had to compete not only with gunpowder, but with nitroglycerine, and an explosion at the Stowmarket plant on 11 August 1871 did not add to the general confidence in its safety. Three magazines containing about 13 tonnes of guncotton blew up, leaving a crater 32 m by 20 m and 3 m deep; an hour later there was a further explosion in the packing house, which killed Edward and William Prentice who were helping with rescue work. Perhaps the Prentice family were unlucky. The incident was generally attributed to internal sabotage, and the rebuilt factory operated, first as the Stowmarket Guncotton Co. Ltd and then as The Explosives Co. Ltd., for thirty-eight years without any further loss of life.

*   *   *

While the case for nitroglycerine was put by Nobel, Sir Frederick Abel spoke for guncotton. The one was an entrepreneur, the other a civil servant, but they were both great chemists, experts in their fields and formidable advocates. Both were also bearded bachelors.

Sir Frederick Abel.

   Frederick Augustus Abel was born in 1827 in Woolwich, where he was destined to spend much of his working life. His grandfather had been the court miniature painter to the Grand Duke of Mecklenburg-Schwerin, and his father was a professional musician. Frederick himself had a lifelong interest in music and was a talented performer, but it was a visit to his uncle, a mineralogist in Hamburg, at the age of fourteen, that kindled his scientific interest. His early schooling was at the Royal Polytechnic, one of the few establishments of the day where the syllabus had any scientific content. And he was clever and lucky enough to be chosen as one of the first twenty-six pupils at the newly opened Royal College of Chemistry, founded upon the recommendation of a committee, chaired by Prince Albert, with the purpose of improving chemical education in England. The college was directed by a famous German chemist, A.W. Hofmann, who had been an assistant at the famous School of Chemistry founded by Liebig in Giessen. Queen Victoria and the prince had been to Germany for the celebration of the 75th anniversary of Beethoven's birth, and they had persuaded Hofmann to come to England for a trial period of two years. In the event, he stayed for twenty years, making very important contributions in all aspects of scientific education.
   Abel was but one of the first of many successful pupils from the Royal College. After various teaching posts he became Professor of Chemistry at the Royal Military Academy at Woolwich and in 1864 he was appointed to the newly created post of Chemist to the Office of Ordnance, which later came to be called

Chief Chemist to the War Office. He occupied that position with great distinction for thirty-four years during which the whole art and practice of gunnery, and the wider general use of explosives, underwent a series of momentous changes, to many of which Abel contributed. His two main achievements were, first, the devising in 1863 of an improved method of manufacturing guncotton and, second, the invention of cordite in 1899, in association with Sir James Dewar, better known as the inventor of the vacuum flask; cordite would eventually replace gunpowder as the main propellant explosive.

In 1881 Abel served on a Royal Commission set up to enquire into the cause of a coal-mine disaster at Seaham; he headed an inquiry into the effects of dust in colliery explosions; and he was a government adviser on the safe storage of petroleum products in the early days of that industry. He was knighted in 1883 for his services to military science, and had a baronetcy conferred upon him in 1893, nine years before his death.

*   *   *

Abel generally gets most of the credit for bringing guncotton into safe, regular use, but von Lenk's insistence on thorough washing was one of the keys to success, and another was suggested in a patent taken out by John Tonkin in 1862. He was the manager of the small, recently founded Cornwall Blasting Powder Co. situated near Truro, which made gunpowder. His patent[13] covered a mixture of gunpowder with guncotton but it was specified that the guncotton had to be pulped before mixing 'in the same manner as is practised by paper-makers, by putting the fibre into a cylinder having knives revolving rapidly working close to fixed knives'. The pulping machine did to fibrous guncotton what a cylinder mower does to grass.

Abel brought von Lenk's and Tonkin's ideas together in his patent[14] of 1865 for 'Improvements in the Preparation and Treatment of Gun-cotton'. He realized that the pulping broke down the fibrous, capillary structure of the guncotton and that this enabled the washing, a process he extended to include repeated boilings, to remove both excess acid and any unstable impurities all the more effectively.

Experimental production began at Waltham Abbey in 1863 and the greater reliability of the guncotton made by the Abel process began to re-establish confidence in the explosive. But the new process was slow and expensive, and Abel was at first disappointed with the reception his achievement received in official circles. This changed in 1871, when a Select Committee reported favourably on the safety of the explosive and recommended the building of a plant at Waltham Abbey to make 250 tonnes per year. The Abel process was also adopted by Messrs Prentice at Stowmarket, and, with some modifications introduced at Ardeer, it lasted until 1905.

One of Abel's assistants, E.A. Brown, discovered that dry guncotton could be detonated very effectively by mercury fulminate, and that the dry guncotton could then be used for detonating wet guncotton. This was useful because wet guncotton could be stored particularly safely, and it could also be compressed into blocks or cylinders as hard as wood. Moreover, the explosive was so stable, in this condition, that it could be cut, turned or drilled by wood-working tools. It could,

A.D. 1865, *20th April.* N° 1102.

Preparing Gun Cotton.

LETTERS PATENT to Frederick Augustus Abel, of the Royal Arsenal, Woolwich, in the County of Kent, for the Invention of "IMPROVEMENTS IN THE PREPARATION AND TREATMENT OF GUN COTTON."

Sealed the 13th October 1865, and dated the 20th April 1865.

PROVISIONAL SPECIFICATION left by the said Frederick Augustus Abel at the Office of the Commissioners of Patents, with his Petition, on the 20th April 1865.

I, FREDERICK AUGUSTUS ABEL, of the Royal Arsenal, Woolwich, in the
5 County of Kent, do hereby declare the nature of the said Invention for "IMPROVEMENTS IN THE PREPARATION AND TREATMENT OF GUN COTTON," to be as follows :—

My Invention has reference to the explosive compound known as gun cotton. Such compound has heretofore been employed either in a loose,
10 fibrous, or woolly state, or of late it has been spun into the form of rovings, yarn, or thread, and has then been formed into cartridges, either by winding, braiding, or weaving.

Now my Invention has for its object to assimilate the physical condition of gun cotton as nearly as possible to that of gunpowder, by mechanically con-
15 verting it into a solid homogeneous state, and imparting to it either a granular or other suitable form that will present the exact amount of surface and compactness required for obtaining a certain rapidity or intensity of combustion.

The method of treating the gun cotton which I prefer to employ in carrying

An extract from Abel's patent application, 1865. (The Patent Office)

therefore, be produced in shapes very suitable for insertion into bombs, mines and torpedo heads, or in standard shapes and sizes for use in quarries and mines.

To make a cheaper, though less powerful, explosive, Abel suggested mixing guncotton with either potassium nitrate or sodium nitrate in what he called 'nitrated guncotton', but the military authorities rejected the products in 1872 in favour of the more powerful wet guncotton. The idea did, however, catch on in mining, quarrying and civil engineering, when George Trench, the manager of the newly formed Cotton Powder Co., which had resurrected the manufacture of guncotton in Faversham in 1873, introduced Tonite, a mixture of approximately equal parts of guncotton and barium nitrate. This was a particularly successful blend which became a serious competitor to dynamite. It was safe in storage and in action, and for many years the railways accepted it for carriage under the same conditions as gunpowder whereas they would not carry dynamite. It was used in constructing the Manchester Ship Canal (1885–94); in widening the Great Western Railway track near Reading in 1891; in building the Severn Tunnel (1873–86); and in rocket distress signals for merchant ships, sound signals for lighthouses and wreck clearance.

Tonite was also about the only form of guncotton used for blasting in the United States. Alfred Victor du Pont had made and tested guncotton as early as 1846, only to conclude that the discovery was brilliant and astonishing but that introducing guncotton into common use was not timely. Nor was the time regarded as ripe when Henry du Pont reassessed guncotton, in collaboration with the American navy, seventeen years later in 1863. DuPonts did not, in fact, make guncotton until 1892, some thirty years after it had begun to be manufactured successfully in Great Britain, and it was always used more widely in Europe than across the Atlantic.

The Abel process for making guncotton was eventually improved upon in 1905, when the so-called Thomson displacement process began to be used. This was simpler, less laborious, cheaper and cleaner than the Abel process, and it provided guncotton both in peacetime and throughout both the First and Second World Wars.

*   *   *

The records show that Schönbein, and others, had hoped that guncotton might serve both as a propellant and as a high explosive but, although it was well established in the latter role by 1875, all attempts to use it to replace gunpowder as a propellant had been very short-lived. Progress in that direction came only when it was realized that guncotton was only one form of nitrocellulose and that another form was just as important. Guncotton was made by prolonged treatment of cotton with hot, concentrated acids; it contained more than 13 per cent nitrogen. By using weaker acids at a lower temperature for a shorter time it was possible to make a form of nitrocellulose containing only about 8–12 per cent nitrogen, which was called collodion cotton, or, simply, collodion. It had been made first at about the same time as guncotton, though the two were not clearly distinguished until later. Collodion was less explosive than guncotton and it had the important difference of being soluble in a mixture of alcohol and ether, unlike guncotton. Collodion is therefore sometimes referred to as soluble nitrocellulose, and guncotton as insoluble nitrocellulose (but that refers only to

Queen Victoria and Princess Beatrice aboard the Admiralty yacht Enchantress, *at the official opening of the Manchester Ship Canal on 21 May 1894. The 58 km long canal, containing five locks, links Manchester to the sea at Eastham, Cheshire. (From* The Illustrated London News, *26 May 1894)*

the mixture of ether and alcohol because both guncotton and collodion are soluble in acetone).

Collodion was first used in photography and in making artificial silk, celluloid, various lacquers and surface coatings, and blasting gelatine. Abel added to that list by using it in 1889 to make cordite, which at long last replaced gunpowder as a propellant and heralded the dawn of yet another era in the explosives saga – that of smokeless powders.

CHAPTER 11

Smokeless Powders

By 1870 nitroglycerine and nitrocellulose products were beginning to replace gunpowder in its role as a high explosive, but as a propellant it still led the field. Not that it was entirely satisfactory, because it fouled the guns, fired irregularly, did not store well, and was dirty and smoky. But there was to be no better alternative until about 1886.

It has always been more difficult to make a good propellant than a good high explosive, just as it is more difficult to sing softly or to dance slowly. The requirements for a good propellant are very subtle and there is a huge variety of guns, both military and sporting, all designed to achieve some particular object, and with different bores, barrel lengths and projectiles. Ideally a propellant must undergo a rapid, but regular and controlled, burning so as to maintain a steady pressure on the projectile while it is in the barrel of the gun. It must not be able to detonate within the barrel or it will shatter or damage the gun; it should be free of smoke and flash and should leave no residue; it should be completely burnt out as the projectile leaves the muzzle; it should be easy to set off, easy to store and to transport; moisture and temperature change should not affect it unduly; and the gases it produces should not be corrosive. All this is a very tall order.

Not all these requirements were, or are, of equal importance. In most cases, the rate of burning is the key issue, which is why nitroglycerine, guncotton and collodion were not satisfactory propellants on their own. They all burnt far too rapidly so that they damaged the gun, while guncotton and collodion also burnt too irregularly owing to their porous, fluffy structure. The resulting high surface area led to a high initial rate of burning but this fell off as the surface got smaller and smaller during the burning. This was counteracted by an increase in the rate of burning as the pressure on the propellant in the gun barrel rose, but with porous materials this rise in pressure was uneven because it was more difficult for the hot gases to escape from within the pores than from the outer surface. As a result the pressure rose more in the interior than at the surface, and the pressure rise could sometimes be high enough to cause the guncotton or the collodion to detonate when used as propellants. Sometimes they would burn reasonably smoothly; sometimes they wouldn't.

It was, then, a matter of breaking down the structures of guncotton and collodion to obtain them in a more rigid form, and this was done by blending them together, or with nitroglycerine. It is surprising, at first sight, that good propellants were made by mixing high explosives, but it was the way in which it was done that mattered.

* * *

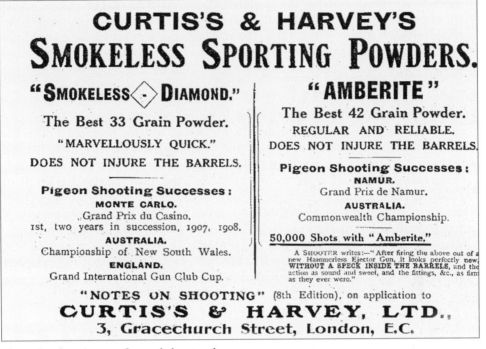

An early advertisement for smokeless powders.

The first smokeless powder which worked satisfactorily in military, rifled guns was made by the French Professor of Chemistry Paul Vielle in 1886. At that time gunpowder was known in France as Poudre N ('poudre noir' or 'black powder') so the new powder, which was white, was called Poudre B ('poudre blanche' or 'white powder'). It was made by kneading a mixture of the two forms of nitrocellulose – guncotton and collodion – with a mixture of ether and alcohol. The collodion dissolved to form a solution which then absorbed the insoluble guncotton. The resulting pasty mass was rolled into sheets which could be cut into squares, or extruded through dies into strips of various shapes, before drying. The fluffy, porous structures of the nitrocelluloses had been broken down, and the product was a horn-like, non-porous material.

This was about three times more powerful than gunpowder and enabled the calibre of a gun to be lowered, with a consequent reduction in the weight of the ammunition. As a result, soldiers could carry more ammunition and what they did carry was more effective at longer ranges. Poudre B was immediately adopted by the French army and navy, partly because it was so successful and also because Vielle had good political connections. It had an immediate effect on the nature of warfare and for a time the French Lebel rifle, with a bore of 8 mm as compared with the 19 mm of Brown Bess, was the predominant infantry weapon in Europe. But, just like other previous explosives, Poudre B was not perfect and there were numerous accidents, notably the complete destruction of two of France's best battleships: the *Jéna* in 1907 and the *Liberté* in 1911. Both blew up in Toulon harbour with great loss of life. Once more, it seemed, a new explosive could not be introduced without exacting a severe toll.

* * *

Shortly after Vielle's discovery of Poudre B, Nobel entered the scene once again with his fourth great invention, ballistite. He seems to have been led to this via celluloid, which had been patented as one of the earliest plastics in 1870, and which was made from a mixture of collodion and camphor dissolved in alcohol. Nobel realized that celluloid would burn, but not rapidly enough for it to function as a propellant, so he replaced some of the camphor with nitroglycerine. The first ballistite consisted, then, of 10 per cent camphor with equal amounts of nitroglycerine and nitrocellulose. Nobel's patent specified that the latter was to be 'of the well-known soluble kind'.[1] It was a phrase he would come to rue. At first benzene was used to mix the materials, but Nobel soon found that the camphor could be omitted and that the nitroglycerine and the collodion would inter-mix simply on warming. The resulting paste could be made, like Poudre B, into a variety of shapes and sizes of a hard material.

Nobel offered the use of his patent rights to l'Administration des Poudres et Salpêtres in France but they were more than satisfied with Poudre B so the offer was rejected. Thinking that this decision had been taken because Vielle had greater political power, Nobel wrote, very scathingly, 'for all governments a weak powder with strong influence is obviously better than a strong powder without this essential complement'.[2] Against this background, Nobel began to make ballistite in Italy, and in 1890 sold his patent rights to the Italian government for half a million lire. This sale, giving military advantage to an unfriendly foreign power, aroused great passions in France and, although Nobel had lived in Paris for seventeen years, he was subjected to much abuse. The press campaigned against him, his laboratory was searched and closed down by the police, he was forbidden to make any more ballistite at his factory in Honfleur, which was mainly making dynamite, and he was even threatened with imprisonment for spying.

To add to Nobel's troubles, his affair with Sofie was coming to an end; the world's press, having mistaken his brother Ludvig's death in 1888 for his own, had printed a premature and very mixed bag of obituaries; and the Panama scandal had broken, with Paul Barbe, one of Nobel's ablest lieutenants, involving him in disastrous financial losses and much ignominy connected with shady dealings during the construction of the canal. Nobel felt that he had never behaved anything but perfectly honourably, but all these untoward events seriously affected his health and led to him leaving Paris for San Remo in Italy, though he continued to refer to Paris as his home in his subsequent patent applications.

* * *

Vielle's and Nobel's discoveries were viewed with great alarm in England, and an Explosives Committee, with Abel as chairman and Dewar as a member, was set up in July 1888. Its terms of reference were 'to investigate new discoveries, especially such as affect the use of military explosives, and to submit to the War Office proposals for the introduction of any technical improvements in the field that the committee could recommend'.[3] Many mixtures were tested and, on request, Nobel sent details of his invention and samples of early ballistite made at Honfleur. The Committee found, as did Nobel, that the volatile camphor evaporated from the ballistite with a resulting deterioration in its properties but, even with that proviso, they considered ballistite to be a better product than Poudre B. They therefore asked Nobel to submit samples of his later type of ballistite, without the camphor, but for some reason he never did so. It was uncharacteristic of him to miss any chance of displaying his wares, and he must have realized that large orders for ballistite would be forthcoming if the Explosives Committee recommended it, but he was going through an awkward time in France.

In the event he missed the boat because during this period Abel and Dewar had been collaborating on producing a smokeless powder of their own, and in 1889 they patented the twenty-eighth mixture they had tested.[4] It contained 58 per cent nitroglycerine, 37 per cent guncotton (insoluble nitrocellulose) and 5 per cent Vaseline. The mixture was made up into a paste, using acetone as a solvent, and then extruded through dies to form thin rods of a rubbery material. These were cut up into lengths and, on drying, the acetone was recovered for reuse. It is curious to note how many of the machines used were based on those involved in making bread, cutting pastry or forming spaghetti or macaroni. The material was originally called 'cord powder' or 'the committee's modification of ballistite' but cordite became its permanent name.

It differed from ballistite in three main ways. It contained Vaseline instead of camphor; it had a higher proportion of nitroglycerine; and it used guncotton (insoluble nitrocellulose) instead of collodion (soluble nitrocellulose), which Nobel had particularly specified. The Vaseline was introduced to lubricate the gun barrel and keep it clean, but it proved, in practice to have an important stabilizing effect on the cordite during storage. The higher amount of nitroglycerine enabled cordite to be fired rather more easily than ballistite. But it was the replacement of the soluble nitrocellulose in ballistite by the insoluble form in Abel's cordite that proved most controversial, because this became the main issue when Nobel claimed that Abel and Dewar had infringed his patent rights. At this point the Swedish and English versions of events tend to diverge. Did Nobel's voluntary and helpful disclosures to the English Explosives Committee materially assist Abel and Dewar to develop cordite? Was the use of insoluble nitrocellulose instead of the soluble variety, as specified by Nobel, a significant and original change? Why had Nobel not sent the Committee a sample of his latest type of ballistite?

Abel and Nobel, despite the former's earlier antagonism to dynamite, had been on reasonably good terms as far as technical matters were concerned, but they were both self-assured experts, and neither would give any ground in the ensuing

Machines for mixing cordite. They had a capacity of 200 litres and were fitted with two four-bladed, horizontal shafts which could be rotated in either direction. (Compair Holman Ltd)

argument. The case therefore had to be resolved in the courts: it was heard first in the Chancery Court in 1892 and then went before the Appeal Court and the House of Lords in 1895. The legal process was very prolonged and the case between such giants – which was, in a sense, a battle between Nobel and the British government – created much interest and some sensation. The meaning of the phrase 'of the well-known soluble kind' in Nobel's patent was hotly disputed, partly because the meaning of the word soluble is imprecise unless it is related to some specific solvent, but in the end all the courts decreed that Abel and Dewar had not infringed any patent rights.

Nobel bitterly resented the verdict and he was highly critical of the whole legal process. His counsel, Mr Fletcher Moulton, who in his youth had been regarded as one of the most brilliant mathematicians in England, was accused of 'bungling management' and was likened to a visitor to a museum 'who had seen the flies and the beetles but not the elephant'.[5] Nor did Nobel find any solace in Lord Justice Keys's remarks that,

It is quite obvious that a dwarf who has been allowed to climb up on the back of a giant can see further than the giant himself. . . . In this case I cannot but sympathise with the holder of the original patent. Mr. Nobel made a great invention, which in theory was something extraordinary, a really great innovation – and then two clever chemists got hold of his specifications for the patent, read them carefully, and, after that, with the aid of their own thorough knowledge of chemistry, discovered that they could use practically the same substances, with a difference as to one of them and produce the same results one by one.[6]

Nobel, faced with paying legal costs, wrote: 'People may say that there's no use crying over spilt milk: nor do I, but there is something in grievous injustice when committed by the State, which very much revolts my feelings. A sane sense of right and wrong should not rise from the mob to the Crown, but ought to spread downwards from the summit. The moral of the whole cordite suit would have been pre-expressed by Hamlet if he had said "that something is rotten in the State of Justice". . . . Just fancy a poor inventor having to spend £22,000 in a "friendly" suit to establish his right. I wonder what the brains are like in the stupid heads that concocted such a monstrous humbuggification!'[7] He vented his feelings still further by writing a parody called *The Patent Bacillus*, in which he satirized justice and bureaucracy against the background of the cordite case.

It is doubtful whether Nobel had quite so much to complain about. His ballistite was chosen for use by many countries, including Italy, Germany, Austria, Sweden and Norway; Poudre B was favoured by France and, in a slightly modified form, by Russia and the United States; and cordite was used throughout the British Empire, Japan and, eventually, Germany.

There was not much, in fact, to choose between the three different powders. Poudre B, which contained only the two different forms of nitrocellulose, came to be known as a single-base powder, whereas ballistite and cordite, containing both nitrocellulose and nitroglycerine were called double-base powders. They were all non-porous, hard, horny or rubbery materials. They were all relatively smokeless because they burnt completely, producing only invisible gases, but because the gases were acidic they all caused some corrosion in gun barrels. They were all extremely versatile because, like gunpowder and particularly brown gunpowder before them, they could be fabricated in so many different shapes: round, tubular or oval lengths; strips, discs or flakes; with or without perforations; and of different sizes. This meant that they could be used as the propellant both in small, short-barrelled pistols and in very large guns. By designing the shape and size of the explosive carefully, the time and rate of burning could be controlled very accurately and, as the factors involved came to be appreciated in more and more detail, many different modifications of smokeless powders were introduced.

* * *

The Explosives Committee was wound up in 1891 and in that year manufacture of cordite began at Waltham Abbey whereupon it was adopted for use in the British services with an immediate advantageous effect on the functioning and design of all guns. The Superintendent at Waltham Abbey had rather grandiose plans for making about 900 tonnes of cordite annually but the government decided that some of the supply should come from private firms. They must have been confident that the verdict in the legal battle between Abel and Nobel would end in favour of cordite because in 1894, before the case was ended, they granted contracts to G. Kynoch & Co. Ltd and the National Explosives Co. Ltd. Both agreed to provide 600 tonnes of cordite over a period of three years. Nobel's factory at Ardeer was distinctly out of favour and, although they were making some cordite for export, they did not get a significant government contract until the start of the Boer War in 1899.

George Kynoch was born in Scotland in 1834. He was 'flamboyant, energetic, likeable, and randy, revelling in wealth as long as he had any, undeterred by pettifogging scruples'.[8] He certainly had an entrepreneurial spirit and between 1862 and 1888 he was involved in a number of activities with a flourishing ammunition business as the centre of his little empire. The firm operated in the Birmingham region, making gunpowder at Worsborough Dale, and detonators, percussion caps, fuses, fog signals and cartridges at Handsworth and Witton. Their products, required for both military and peaceful uses, were sold all over the world in keen competition with Eley Bros Ltd, which had been founded in 1828 and which made similar items at factories in Edmonton, London, and Faversham in Kent.

George Kynoch turned his business into a company in 1884 and became its managing director. Two years later he was elected as Conservative Member of Parliament for Aston, and his star was very much in the ascendant. But he was nothing if not an individualist and he fell out with the board of the company in 1888 and was forced to resign. Thereafter the company was directed by Arthur Chamberlain, a proud, honest, austere nonconformist from the same clan as Joseph, Austen and Neville, who did not have much in common with George Kynoch. It was Chamberlain, however, who negotiated the contract to supply cordite even though Kynochs had no facilities at all for making it and at first they had to buy cordite paste from Nobel's. But two new factories were set up, one at Arklow in Ireland in 1895 and another at Kynochtown in Essex in 1897, and the company was reorganized and renamed Kynoch Ltd. It soon began to make blasting explosives as well as cordite and became a potentially serious competitor to Nobel's Explosives.

The National Explosives Co. Ltd, founded in 1887, grew out of the old Kennall Vale Gunpowder Co., which had for some time been trying to become less dependent on gunpowder. New buildings were erected on the north coast of Cornwall between Hayle and St Ives, with the original intention of making dynamite to try to relieve the stranglehold that Nobel's Explosives had on the supply of that explosive to the Cornish mines. The factories were built under the supervision of Oscar Guttmann, who had been brought in as a director of the new company. He was born in Hungary in 1855 and after training as an engineer he specialized in the field of explosives, rapidly building up a reputation both as a writer and a manager. By the age of twenty-three he was editing two mining journals and he was involved with a number of European explosives manufacturers over the next ten years. He had visited England in 1883, and what he saw persuaded him to try his hand as a consultant in London. He arrived in 1887, becoming a naturalized citizen in 1894. In the twenty-three years before he was killed in a car accident in Brussels in 1910, he contributed much to the development of the explosives industry in Britain.

The National Explosives Company was enlarged in 1894, 1899 and 1901, becoming one of the most important manufacturers in Britain, and certainly the largest in Cornwall. Shortly after the turn of the century it was capable of manufacturing 1,000 tonnes of cordite each year, together with 2,000 tonnes of dynamite, gelignite, blasting gelatine and a variety of specialized permitted explosives.

Kynochs, Eleys and National all expanded to meet the increased demands during the First World War but their experiences after 1918, when they could no longer rely on huge military orders, were very different. They all became part of the newly formed Explosives Trades group at the end of the war, but the National Explosives Co. went into voluntary liquidation only a year later, despite strenuous efforts by the staff to keep it going in some form or other in an area where alternative employment prospects were poor. The factory closed in 1920, with only a few workers remaining to unpack and burn the vast stocks of unused cordite.

Kynochs survived, but only at the expense of Eley Bros. When Explosives Trades became Nobel Industries in 1920, Nobel's Explosives was easily the largest shareholder with Kynochs in second place, and Eley Bros in fourth. Third place was occupied by Curtis's & Harvey's who had for some years been buying up small explosives manufacturers. Kynochs were determined to prove that 'they could produce sporting cartridges more efficiently and more cheaply than Eleys',[9] and their factory at Witton was re-equipped between 1920 and 1930 so that it was capable of making cartridges at a rate of one million per week. Eley Bros could not compete and went into liquidation in 1928, with only the famous name remaining. Meanwhile Kynochs had been forced to give up manufacturing explosives, which entailed shutting the plants at Arklow and Kynochtown, but in yet another reorganization it became the major constituent of the new ICI Metals Group formed in 1927.

*   *   *

In response to news of the European advances in making smokeless powders to replace gunpowder, the United States both expanded its naval and military research effort and encouraged DuPonts to enter the field.

Much of the early progress was directed by Charles E. Munroe, one of the founding members of the American Chemical Society. He was born in East Cambridge, Massachusetts, on 24 May 1849 and, after graduating in chemistry, he taught at Harvard until 1874 when he was appointed Professor of Chemistry at the US Naval Academy in Annapolis. It was there that he turned his attention to explosives and he was recognized as an authority in the field by the time he moved to the Naval Torpedo Station at Newport, Rhode Island, in 1886. He began experimenting with smokeless powders in 1889 and by 1891 he had

Charles E. Munroe. (The Hagley Museum and Library)

patented a material not unlike cordite or ballistite, which he called Indurite but which was popularly known as Naval Smokeless Powder. Munroe became Professor of Chemistry at Columbian College (later George Washington University) in Washington in 1892. He retired from that post in 1918, but remained active in many spheres being, for example, the Chief Explosives Chemist of the Bureau of Mines from 1919 to 1933. He died in 1938, at the age of eighty-nine, by which time he had become something of a legend in the history of explosives in the United States.

Around the time that Munroe began his work on smokeless powders, DuPonts had been persuaded to send a representative to Europe to try to buy up any patent rights that might be available. When he came back somewhat empty-handed, the company decided that it would have to forge its own way ahead and to that end built a new centre for research and production at Carney's Point, New Jersey, in 1890. As times were peaceful, the company concentrated on developing smokeless powders that could be used for sporting purposes and by 1893 they were manufacturing such a product, but the requirements of the services were left mainly in the hands of government establishments. It was a period of experimentation and the Navy in particular did some extremely good research work but the outbreak of the Spanish-American war in April 1898 found the

American forces distinctly short of adequate supplies of good quality smokeless powders.

They had to rely on gunpowder, particularly on brown or cocoa powder, which had been extensively developed by DuPonts in collaboration with the Army and Navy, but they did have one ship – the USS *New Orleans* in which cordite from Europe was being tried out. Another experiment, which to modern eyes seems very bizarre, was also attempted. One night in April 1889 the USS *Vesuvius* nosed her way to within 1.6 km of the coast off Santiago de Cuba and fired some shells, each containing 225 kg of dynamite, from three guns mounted in her bows. They fell short of the target, Morro Castle, and did little damage but it was an event unique in naval warfare because the shells were fired from pneumatic guns operated by large air compressors. Similar guns, with a range of 3.2 km, were mounted on some coastal defence batteries. They were all very unwieldy and highly inaccurate and the whole development, which had taken a number of years, was abortive.

After the war the United States moved rapidly away from gunpowder and even more rapidly from pneumatic guns. Many of the naval and military patents were shared with DuPonts, at little or no cost, and they quickly became the nation's major manufacturer of military smokeless powders. So much so, that at the time of the Sherman Act legislation in 1907 they were the only civilian firm making them. This was one of the most contentious issues involved in the case against DuPonts but in the end they were allowed to maintain their monopoly position on the grounds that collaboration between them and the services had always been beneficial to the nation. To introduce some measure of competition, a Naval Powder Factory had been built at Indian Head, Maryland, and an Army Ordnance Factory at Picatinny Arsenal in Dover, New Jersey. These arrangements certainly enabled the United States to produce vast quantities of smokeless powders during the First World War.

* * *

The build-up of the smokeless powder industry in America coincided with a significant change in the composition of cordite in Great Britain. Experiences during the Boer War between 1899 and 1902 proved beyond doubt that cordite was a big improvement on gunpowder, but it also brought to light some drawbacks. In particular it was found that cordite caused very serious corrosion in gun barrels and this led to the composition being changed to 65 parts guncotton, 30 parts nitroglycerine and 5 parts Vaseline. This modified form was called Cordite MD (the MD from MoDified) and the original form came to be known as Cordite Mark I.

Cordite MD was the form used at the start of the First World War but by 1916 its production in the required quantities was threatened by a shortage of acetone. At that time home production of acetone depended on heating wood but it was inadequate both because of the shortage of timber in this country and because it took a great deal of wood to make a little acetone. Aeroplane manufacturers also required large amounts of acetone for making varnishes. As a result, much acetone had to be imported from the United States and it became difficult to ensure satisfactory supplies from that source. It is ironic that the damage was in some ways self-inflicted. Had an earlier British government opted to use ballistite instead of cordite all might have been well because acetone was not needed in making ballistite.

Fortunately, the matter was resolved by the intervention of Chaim Weizmann, the third of fifteen children of a struggling timber merchant who lived near Pinsk, in Western Russia, in a poor area set aside for Jews and known as the Pale of Settlement. Chaim received his early education in local Russian schools but, as Jews were not encouraged to enter Russian universities, he moved westwards, at the age of eighteen, to continue his studies in Germany and later in Switzerland. He came to England in 1904 to do research at Manchester University and, towards the start of the First World War, while investigating a new method of making synthetic rubber, he discovered on an ear of corn a bacillus that could convert starchy materials, such as maize, into acetone. The process was similar to the age-old operation of using yeast and other enzymes to ferment starchy substances into alcohol. The new bacillus was called *Clostridium acetobutylicum* or BY (bacillus Weizmann) for short.

Winston Churchill, the First Lord of the Admiralty, was at the time urgently seeking a source of 30,000 tonnes of acetone to ensure adequate supplies of cordite, and when he heard of the new process, early in 1915, he gave its discoverer every possible encouragement to develop it. Further support came shortly after from Lloyd George, the newly appointed Minister for Munitions, to whom Weizmann was introduced by C.P. Scott, the editor of the *Manchester Guardian* and one of Lloyd George's most trusted political advisers.

By the start of 1916, following intensive research, mainly at the Nicholson gin factory in Bow, it became possible to operate the new process successfully on an industrial scale, and production of acetone began in six British distilleries which were commandeered for the purpose. At one point the whole enterprise was threatened when it became difficult to maintain adequate supplies of maize, but the need for acetone was so important that a national collection of horse-chestnuts was organized to provide an alternative supply of starchy material.

Weizmann's success in solving the 'acetone problem' had very important political repercussions. He had always been an ardent Zionist, and his outstanding war work helped to ensure that his pleas on behalf of his people, towards the end of the war, did not fall on deaf ears. He was introduced to Mr Balfour, the Foreign Secretary, and there can be no doubt that his influence and advocacy had a great deal to do with the eventual Balfour declaration in 1917, which, subject to certain provisos, viewed with favour 'the establishment in Palestine of a national home for the Jewish people'. When the State of Israel was eventually set up in 1948, Weizmann, at the age of seventy-four, was appointed as its first President. The wild dreams which he had had in his boyhood days in Russia that the Jews would, one day, 'rise again in their ancient land' had come true. But he cannot possibly have dreamt that acetone or cordite would have anything to do with it.

To maintain supplies of cordite while the new method of making acetone was being built up, a third modification, called Cordite RDB (Research Department formula B), which was very much like ballistite, was introduced. It contained 52 parts collodion, 42 parts nitroglycerine and 6 parts Vaseline. It had the all-important advantage that it could be made using an ether-alcohol solvent instead of acetone, but it was only a temporary expedient because it did not store very well. Further, more sophisticated, refinements followed over the years. It became possible to lengthen the storage life by adding stabilizers; cordite could be made without using any solvent; the burning characteristics could be altered by various surface treatments or by adding moderants or deadeners; and, where necessary, the flash from the powder could be decreased. The material is so versatile that twenty modifications are included in the modern Health and Safety Executive list of Authorized Explosives. This smokeless powder par excellence served nobly in two world wars and is still in use today. Gunpowder may have been a jack-of-all-trades but cordite was master in the propellant business.

* * *

Smokeless powders used in military small-arms are also suitable, with little modification, for sporting rifles, but something different is needed for shotguns. The larger bore, the relatively low muzzle velocity, the small weight of the shot, and the light construction at the muzzle end of the gun, so that it can be handled easily, all demand a powder which both ignites and burns very rapidly without building up too high a pressure. The longer the delay between pulling the trigger and the shot leaving the muzzle of the gun, the less accurate the shooting. To make something that meets all the subtle requirements taxes all the skills of the powder maker even today.

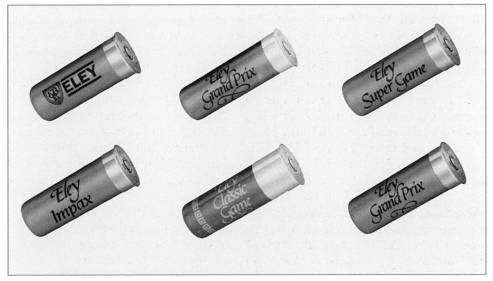

A selection of shotgun cartridges. (Eley Hawk Ltd)

The first powder to prove useful in shotguns, though that was not the original intention, was made by Major Schultze, a Prussian artillery officer, in 1864, some twenty years before Poudre B, ballistite or cordite. He purified wood, to some extent, by boiling and bleaching it and then nitrated it to make an impure nitrocellulose mixture. It was mixed with potassium and barium nitrates to make Schultze powder. This was patented in England in 1864 and was manufactured for many years by the Schultze Powder Co. Ltd at a factory at Eyeworth in the New Forest in Hampshire. An improved mixture, which could be used in rifles as well as shotguns, was made in Austria under the name of Collodin, but it only survived until 1875 when the government banned it because its success seemed likely to threaten their gunpowder monopoly. An English mixture called E.C. Powder followed in 1882. This was made by the Explosives Company at Stowmarket, a successor to Thomas Prentice & Co., and consisted of a mixture of pulped guncotton with potassium and barium nitrates. The components were mixed, while the guncotton was wet, to form a paste that could be converted into granules of various sizes by squeezing it through sieves. The granules were then treated with a solvent to harden the surface, and finally dried. Manufacture of both Schultze and E.C. Powders began in the United States in the 1890s.

Since then many other similar so-called bulk powders have been made by a

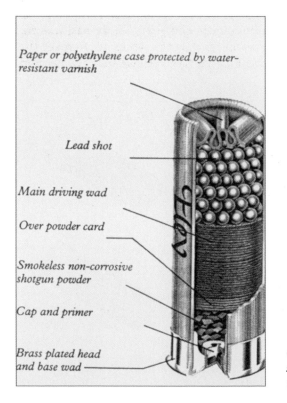

Paper or polyethylene case protected by water-resistant varnish

Lead shot

Main driving wad

Over powder card

Smokeless non-corrosive shotgun powder

Cap and primer

Brass plated head and base wad

Cross-section of an Eley Grand Prix shotgun cartridge for game-shooting. (Eley Hawk Ltd)

number of different manufacturers and with a wide range of additives and surface treatments to achieve some particular purpose. Generations of sportsmen will be familiar with the trade name of their particular, favourite cartridge. In one of the more significant newer processes, introduced in the United States in the 1930s, solvents that were immiscible with water were used to dissolve the powder mixture into a viscous syrup, which, on agitation with water, split up into spherical granules. The suspension was then warmed to evaporate off the solvent, leaving solid spheres that could be separated according to size and dried. The product was known as a ball–grain powder.

*   *   *

The most recent application of smokeless powders has been in the motors of missiles and space rockets when it is preferable to employ a solid rather than a liquid mixture. The propellants used are either of the cordite type, or they are what are known as composite propellants. The latter contain an oxidizer such as ammonium nitrate or ammonium perchlorate, together with a plastic, such as polyvinylchloride or polyisobutene, which can act both as a fuel and a binder. Aluminium may also be added when it is necessary to raise the burning temperature. Other uses include the rapid starting of jet engines in fighter aircraft and the operation of safety and security alarms by electric motors which can be switched on very reliably and very quickly by the firing of a smokeless powder.

It is all a far cry from the powder which Schultze made over 120 years ago, even though his basic recipe has not been greatly changed. There has, indeed, been some move towards using wood, as Schultze did, or even straw, to make the necessary nitrocellulose for modern smokeless powders because they are cheaper than cotton.

CHAPTER 12

Lyddite and TNT

In May 1915 Mr Asquith was Prime Minister of a Liberal government in the United Kingdom; Lord Kitchener of Khartoum was the Secretary of State for War; Sir John French was the Commander-in-Chief of the British forces fighting the Germans in Europe; and Lord Northcliffe owned both *The Times* and the *Daily Mail*.

The First World War had been under way for nine months, and many of those who had answered Kitchener's well-advertised call to arms, 'Your Country Needs You', were on their way to France, but there had been little or no British success on the battlefields. More shells had been fired in a 35-minute barrage at the start of the battle of Neuve Chapelle than in the whole of the Boer War – but it had cost 11,200 casualties to gain a small salient of ground. Things were not going well in the Dardanelles campaign, and Haber had just organized the first German gas attack at Ypres.

On 9 May Sir John French launched an attack around Festubert to try to capture Aubers Ridge, but the battle lasted for only one day; little was achieved, and 481 officers and 11,161 men were killed. This was the final straw for Sir John. He had repeatedly, and angrily, demanded better supplies for his armies from Lord Kitchener and of the latest battle he wrote: 'After all our demands, less than 8 per cent of our shells were high explosive, and we had only sufficient supply for about 40 minutes of artillery preparation for this attack . . . As I watched the Aubers Ridge, I clearly saw the great inequality of the artillery duels, and, as attack after attack failed, I could see the absence of sufficient artillery support was doubling and trebling our losses in men.'[1]

But he went much further. He reported the situation direct to the Cabinet and, for good measure, he briefed Colonel Repington, the military correspondent of *The Times*. The headline in that paper, on 14 May, ran: NEED FOR SHELLS; BRITISH ATTACKS CHECKED; LIMITED SUPPLY THE CAUSE, and the article laid bare the facts. The shattering impact of that report on the public was reinforced by letters[2] from soldiers at the front. 'We have just to sit tight in the trenches while the German high explosives shatter them to bits.' 'It makes your heart break to see those men going forward and then held up – one after another, fighting and struggling, until, wearied out, they collapse like a wet cloth. And why? Because there is not an adequate supply of high explosives to blow the wire to bits and let our men get at the enemy.'

Something had to give, and by the time the *Daily Mail* was running its 21 May

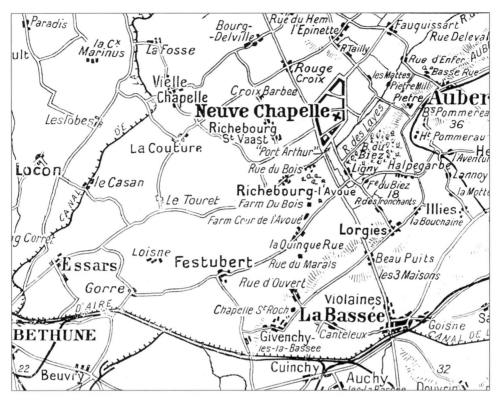

Many battles were fought in this area of north-western France during the First World War.
(From The Times History of the War, *vol. V, 1915)*

headline, THE SHELLS SCANDAL; LORD KITCHENER'S TRAGIC BLUNDER, Mr Asquith was busy reconstructing his government. Give us the guns and we will finish the job was the new order of the day.

* * *

Gunpowder had been replaced by guncotton as the filling in military mines, torpedoes and demolition charges around 1870, following the recommendation of a committee of Royal Engineers, and superseded as a propellant in guns by cordite around 1890. But it remained as a shell filling until 1896 because both guncotton and nitroglycerine mixtures were too sensitive to be satisfactory replacements. Shells filled with them were liable to be detonated in the gun barrel by the shock to which they were subjected on firing and, if that happened, the gun was severely damaged.

Shell-fillings must be powerful but not too sensitive. They also have to store well under all sorts of conditions; they have to be able to withstand rough handling and transport; they should have a low melting point so that they can be poured into a variety of containers; they should not react with the metals of which the containers are made; they should have a low density so that the shell can be as light as possible to give it the longest range; and they should produce some smoke for observation purposes when they explode on hitting the target. So that they can be stored safely in dumps, it is also important that they should have no tendency to detonate sympathetically. That is, a dump of shells should be able to withstand the impact of an exploding shell without the whole dump blowing up.

Picric acid, discovered in 1771, was the first material that began to meet the requirements. It is a bright yellow solid, with a melting point of 122°C and a very bitter, poisonous taste; hence its name from the Greek *pikros* meaning bitter. It was the first artificial dye to be used, in 1799, for dyeing silk and wool, and for many years it was used for treating burns until better procedures were introduced during the Second World War. Sprengel discovered in 1871 that picric acid could be exploded by a sufficiently powerful detonator and tests were carried out by Messrs John Hall & Sons at Faversham, but it was not taken up as a useful explosive until it was adopted as a filling for French shells in 1885. The product used, a mixture of picric acid with some collodion, was called Melinite. Together with the introduction of Poudre B at the same time, it meant that the French had, temporarily, both the best propellant and the best high explosive for military use.

Abel's Explosives Committee, set up in 1888, was very interested in these developments, particularly as Abel had himself suggested a mixture of 2 parts ammonium picrate and 3 parts potassium nitrate as an explosive, called picric powder, as early as 1874. Tests on Melinite were carried out at Lydd in Hampshire and these were so successful that picric acid, under the name of Lyddite, became the principal shell filling in the United Kingdom, with Abel's picric powder being used as a booster or primer to assist in its detonation. Many other countries followed suit with picric acid being used as Granatfüllung 88 in Germany, as Pertite in Italy, as Shimosite in Japan, as Coronite in Sweden, as Picrinite in Spain and as Ercasite in Austria.

Lyddite-filled shells were first used by British forces at Omdurman in 1898 and in the Boer War a year later. At the battle of Omdurman, 50,000 dervishes, armed mainly with spears, chose to attack a force of 20,000 British and Egyptian soldiers, backed up by forty-four pieces of artillery and a flotilla of naval gunboats, under the command of General Kitchener. Within an hour or two, the power of the British explosives had killed over 11,000 dervishes and the Sudan had been recaptured with the loss of 48 British and Egyptian lives. Kitchener said that he thought the enemy had been given a 'good dusting' and a war correspondent wrote that 'it was not a battle but an execution'.

But Lyddite-filled shells were less successful during the Boer War. The concentrated blast of the shells on landing made a lot of noise and sent up a cloud of rock and earth but inflicted little damage on Botha's well-emplaced guns and well-entrenched riflemen. In these conditions, it was the Shrapnel shell, invented by Lieutenant (later General) Shrapnel and first used in 1808 against the French, that caused the most damage. Captain Mercer had realized, during the siege of Gibraltar between 1779 and 1783, that a shell exploding in the air, over the heads of enemy troops, could do more damage than one exploding on the ground. To achieve this he had adopted the makeshift solution of shortening the timing of the fuse on a normal shell. Shrapnel developed that idea by designing a special shell with a thin casing filled with heavy, metal bullets and containing a small charge of gunpowder just big enough to open up the case. When the shell was fired it was fused to burst open above its target so as to eject a shower of lethal bullets. The original shell was called spherical case shot and it was not entirely satisfactory because of difficulties with the time fuse which didn't always function reliably. But the problems had largely been overcome by 1852 when the shell was renamed the Shrapnel shell in honour of its inventor. The design of the shell has become much more sophisticated over the years but the simple idea on which it is based has not changed, and the shell has been so effective against personnel ever since it was first used that shrapnel is now a household word.

*　*　*

Picric acid, or, to give it its full chemical name, trinitrophenol, had been made on a laboratory scale in 1841 in a reaction involving phenol, concentrated sulphuric acid and concentrated nitric acid, but the British found it difficult to scale this up on to a commercial scale at the end of the century because of the backward state of their chemical industry. Initially, phenol was obtained from coal-tar but when supplies from that source could no longer meet the demand, alternative methods of making it from benzene were developed in some haste and some was imported. There were also shortages of sulphuric and nitric acids.

Despite these problems Lyddite remained the main military high explosive in Britain well into the First World War. But it was by no means ideal and it was inferior to the trinitrotoluene, TNT, adopted by Germany in 1902. It was very difficult to detonate, particularly if wet, so that it frequently misfired; whole dumps of Lyddite-filled shells were liable to blow up if hit by one hostile shell; it produced very black smoke when it exploded which made air observation on possible targets difficult; and it could not be melted by steam or hot water, which was the safest way of obtaining an explosive in its liquid form for pouring into shells. Nor was it particularly safe: in its early days it was involved in at least six accidents, including one at Woolwich Arsenal in 1903, when sixteen people were killed and fourteen injured. But it had two more important defects. First, because it was an acid it reacted with metals to form picrates and these were liable to explode when subjected to shock or friction, detonating the remaining Lyddite. To try to prevent this, all metal surfaces within a Lyddite-filled shell had to be given an acid-resistant coating by being treated with copal varnish. Secondly, instead of penetrating armour-plating, which was necessary to inflict the maximum amount of damage, Lyddite-filled shells exploded on impact with the surface of the armour because the Lyddite could not withstand the shock. This deficiency was highlighted at the battle of Jutland in 1916, in which 155 ships were involved. Both sides claimed victory, but the British Navy, with superior gun-power, lost more ships, including such famous vessels as *Indefatigable*, *Queen Mary*, *Invincible*, *Black Prince*, *Warrior* and *Defence*, and twice as many men as the Germans. There have been endless arguments about the naval tactics employed

The sinking of HMS Invincible *at the battle of Jutland, 31 May 1916. An enemy shell entered the starboard turret and detonated 50 tonnes of cordite in the magazine below. The explosion broke the ship in half and there were only 6 survivors out of a crew of 1,031. (Imperial War Museum, SP 2468)*

by both admirals but there can be no argument about the superiority of the German shells, which were filled with desensitized TNT and not with Lyddite, and about the greater strength of the German armour-plate.

It was a marked example of how even a slight technical superiority can tip the scales in warfare. It is not simply a matter of getting there fastest with the mostest; it must also be the best. Indeed, modern warfare, particularly when the tank entered the fray, became more and more a battle between the strength of one side's armour-plating and the power of the adversary's armour-piercing shells, which depended both on their design and on their filling. The replacement of Lyddite, in Britain, by the better TNT, already under way before 1916, was accelerated by the result of the battle of Jutland.

* * *

That Germany was well ahead of Britain in making high explosives around 1900 was another example of the old story of the difficulty of making technological change in Britain. The German chemical industry was far more advanced and its greater strength and adaptability was well exemplified by the synthetic dyestuffs branch. A British chemist, W.H. Perkin, had discovered a new aniline purple, or mauve, dye in 1856, and Fullers of Perth, a famous firm of dyers, had rated it as 'one of the most valuable that has come out for a very long time'. The dye was originally manufactured by Perkin, with help from his father and brother, at a factory near Harrow, and it was immediately very successful. Its colour can still be seen on early penny stamps. Yet by 1900 it was Germany that was the centre of the world dyestuffs industry, and in 1914 Britain was importing 80 per cent of the dyes it required. Even the dyeing of military uniforms and socks presented severe problems. The easier availability of raw materials in Germany was certainly a factor in their success, but they also had a better system of technical education, which was heavily subsidized by the State. It was also quite normal for the top jobs in German firms to be occupied by gifted technical people. That was not so in Britain.

So it was that German attempts to find something better than Lyddite were more successful than those in Britain. Willbrand had made a chemical called trinitrotoluene in 1863, and this had been adopted as a shell-filling in 1902. The value of what became known as TNT or Trotyl was not realized in Britain until just before the First World War, even though samples of the new material had been tested in 1902 and again in 1905. Alas, it was rejected on the grounds that it lacked power and was difficult to detonate, so it was left on the shelf as just another research project.

TNT, like picric acid, is a yellow solid. On the debit side it is less powerful, harder to detonate, and more difficult to make than Lyddite, but in credit, it is cheaper, less poisonous, completely unaffected by dampness, and less dense. Moreover, it has a melting point below 100°C so that it can be melted by hot water or steam which makes shell-filling safer and easier; it is not acidic so that it does not react with metals; and the difficulty in detonating it makes it so stable that in 1910 it was exempted from the provisions of the 1875 Explosives Act, not being regarded as an explosive so far as manufacture and storage were concerned. As it is less sensitive to shock than Lyddite, it can be used (particularly with additives to desensitize it still further) as a filling for armour-piercing shells which will not explode simply on contact with an armoured surface.

The difficulty in detonating TNT was overcome by using stronger boosters than those required by Lyddite, but the main problems of introducing TNT into the British services were those of decision-making and of learning how to manufacture it as readily as the Germans could. It was made from toluene, concentrated sulphuric acid and concentrated nitric acid, and all three raw materials were more readily available in Germany than in Britain. The Haber process for making ammonia, and hence nitric acid, originated in Germany, and, though the Contact process for making sulphuric acid was originally a British discovery, it was exploited most energetically in Germany to meet the requirements of their dyestuffs industry. Toluene, like phenol, was mainly obtained from coal-tar but it needed 600 tonnes of coal to provide 1 tonne of toluene and the German plants were more advanced than those in Britain, again because of the requirements of the dyestuffs manufacturers. When the need for TNT was finally realized in Britain in 1913, the Germans were already eleven years ahead and Britain had certainly not heeded the 'Be Prepared' motto adopted by Baden-Powell for the Boy Scout movement which he founded in 1908.

* * *

At the outbreak of the First World War on 4 August 1914, Britain had few resources with which to wage a prolonged land campaign. The Navy, as the traditional defence force and the guardian of the sea-lanes, was in good shape, but the professional Army, excellent for policing the empire, was far too small. Above all, there was a terrifying lack of equipment, so that the volunteers, superbly recruited in their thousands by Lord Kitchener, had to train and fight almost empty-handed in the first few months while the existing system was completely overhauled.

This was no easy task. Everything was under the rigid control of the War Office, which was weighed down by bureaucracy, tradition, indecisiveness, distrust of business men, lack of imagination, and a dread of taking responsibility. Though Lord Kitchener was a sapper, most of the generals were cavalry men, still wedded to the horse and to their recent experiences in the Boer War. Lloyd George, a British Cabinet minister but no lover of top brass, said of them that their 'military imagination makes up in retentiveness what it misses in agility'.[3] He was equally damning of their French counterparts. To deal with them, he wrote, 'was to contend not with a profession but with a priesthood, devoted to its chosen idol'.[4]

In the first few months of the war the Germans had complete control. The concrete fortifications of Liège, in Belgium, thought to be well-nigh impregnable, were shattered by high explosives within a few hours; the fortress of Namur, which *The Times* described as a 'tough nut which might be expected to take even Prussian jaws six months to crack' fell in two days around 22 August; by 9 October Antwerp had been captured by a force three times smaller than its garrison; and the Allied armies were rolled back remorselessly to the long lines of the Western Front on which they were to be penned for so long. The only reply to the battering handed out by the giant German shells from 42 cm howitzers – nicknamed 'Big Berthas' – came from the British 23.4 cm howitzer, fresh from its recent trials and nicknamed 'Mother'. Yet the British War Office was still putting its trust in the shrapnel shells which had done so well in the very different conditions of the Boer War. Their impact on German trenches was summed up by an officer writing from the front as 'less useful than rain, not being wet'.[5]

Something had to be conjured up out of the blue but the existing institutions were ill-equipped for legerdemain. The War Department was oblivious to new ideas and methods. A General War Committee had been set up by the Royal Society in November 1914 to 'organise assistance to the Government in conducting or suggesting scientific investigation in relation to the war',[6] and this

Big Bertha. (Imperial War Museum, Q65817)

was used considerably by the Admiralty but was shunned by the War Department. The government establishments, where action was centred, were not well organized. The Royal Arsenal at Woolwich sprawled over about 23 sq. km, but the 240 km of internal railway was quite incapable of effectively moving supplies in, out or around. There was little contact with outside industrialists and it was run, in the opinion of the British Minister of Munitions, by people 'who were entrenched in well-worn traditions behind entanglements of red-tape, and all ready, from Alpha to Omega, to die in their ditch rather than surrender the fortress held by them and their official forefathers to the barbarians who threatened their empire from the dark forest of politics'.[7] His French counterpart, after a visit, called it 'une vielle boîte'.[8]

Such was the background out of which the shell scandal blew up, triggered by the headlines of *The Times* and the *Daily Mail*, and against which the government fell. On 26 May 1915 a coalition of Liberals and Conservatives replaced the old Liberal government, with Mr Asquith still acting as Prime Minister. One of his first acts was to create a new department, the Ministry of Munitions, and to

A cartoon showing Lloyd George leading the charge to improve the supply of munitions. (From Punch, *21 April 1915)*

appoint Lloyd George, who had been Chancellor of the Exchequer, as its first minister. The main credit for the realization that the war could only be won if the chemists, engineers and manufacturers could produce enough of the right equipment quickly must go to Lloyd George, even though his critics would say that he changed his mind from the position he had adopted at the Treasury. However that may be, a change of mind was what was needed, and Lloyd George set about his gigantic task in an old, mirror-panelled drawing room in 6 Whitehall Gardens, formerly the home of a well-known art dealer. He took stock with the help of two secretaries, two tables and one armchair.

Within two months he had collected a team of ninety talented and experienced men from all walks of life, including Mr Fletcher Moulton, who in the 1890s had acted as counsel for Alfred Nobel in his legal action for patent infringement against Abel. Now, as Lord Fletcher Moulton, an eminent judge, he came to head the Committee on High Explosives. The various teams toured the industrial areas of the country, set up new organizations and procedures, dealt with labour and trade union issues, introduced women workers, legislated to prevent any exploitation of profits, arranged for the release of key personnel from the services, extended the manufacture of machine tools, and continually upheld the morale of the new workforce by arranging royal visits and introducing good welfare and housing schemes. They even went so far as to control the sale of alcohol when it became clear that increased consumption was affecting production (so much so that the number of gallons of alcohol sold each year fell from 89 million in 1914 to 38 million in 1918).

By the end of the year, the royal factories at Woolwich, Waltham Abbey, Enfield Lock and Farnborough, which had been transferred to the War Office in the autumn, had been supplemented by seventy-three new national factories that opened all over the country. Of these, 8 were for making explosives, 49 for making shell casings, 14 for shell-filling and 2 for making gauges. By the end of the war there were 218 factories in all.

*   *   *

The production of empty shell cases rose very dramatically because it only involved fairly traditional engineering processes, but it was less easy to fill them with explosives which were in distinctly short supply. At the start of the war the Royal Gunpowder Factory at Waltham Abbey was producing 68 tonnes of gunpowder and cordite each week, and there was some production in private firms. Some TNT had been made at Ardeer since 1907 but the total amount of high explosive available would have been laughable if the situation were not so serious. While the Germans were actually firing off about 2,500 tonnes of TNT every week, the total production of both TNT and Lyddite in Britain was less than 20 tonnes per week. TNT had only recently been adopted as the principal high explosive and the production of Lyddite was being run down because the War Office had let it be known that it was to be superseded.

Immediately after the passing of the Defence of the Realm (Consolidation) Act on 27 November 1914, which gave the government power to take over factories, the Rainham Chemical Works on the Thames opposite Woolwich had been commandeered and was used to purify the few tonnes of TNT that were available. By January 1915 the first national factory, organized by Lord Moulton, was being built at Oldbury by Messrs Chance & Hunt. A big factory followed at Queen's Ferry for making guncotton, and one at Gretna in Scotland for making cordite. Only 440 tonnes of Lyddite and TNT were delivered, by Lord Moulton's department, in 1914 but this figure rose to 114,865 tonnes in 1916. Delivery of propellant explosives was 5,382 tonnes in 1914, rising to 208,085 tonnes in 1916. By the end of the war the total production had exceeded 275,000 tonnes of Lyddite and TNT, and 460,000 tonnes of propellant explosives.

All that effort would have been to little avail had it not been for help from across the Atlantic. The United States, like Great Britain, had been slow to begin manufacturing TNT, but within a few weeks of the outbreak of war, DuPonts had delivered very large quantities of it, and other explosives, to Great Britain, France and Russia. The value of this aid was acknowledged by Lord Moulton when he said that the British and French armies could not have held out during the early months of 1915 without it.

By the end of the war in 1918, DuPonts had shipped about 635,000 tonnes of military explosives to the Allies. They had also supplied over 355,000 tonnes of commercial explosives to their home industries, but when the United States entered the war on 16 April 1917 they even managed to step into a still higher gear to meet the vast, new demands of the American military machine. To do so, the company had expanded its workforce from around 6,000 just before the war started to a peak of 100,000, of whom, alas, 350 lost their lives in accidents.

Very considerable amounts of explosives also came from other American manufacturers and from Canada. South Africa also contributed in the shape of Kenneth Quinan, the nephew of W.R. Quinan, who was seconded at a moment's notice from his job as general manager of the Cape Explosives Works to become Lord Moulton's right-hand man. So valuable was his assistance that he was made a Companion of Honour and *The Times* wrote that 'it would be hard to point at anyone who did more to win the 1914–18 War for Britain than K.B. Quinan'.[9]

* * *

The tonnage of TNT actually manufactured was made to go as far as possible by mixing it with ammonium nitrate. This mixture had been used by the French under the name of Schneiderite, but the British composition was called Amatol. It was made simply by mixing warm, granulated ammonium nitrate with molten TNT to give different grades containing 40, 50 or 80 per cent of ammonium nitrate. The Amatols were much cheaper than pure TNT because ammonium nitrate was relatively cheap, and they were only inferior in so far as they were affected by damp because ammonium nitrate is hygroscopic, and because the mixture containing 80 per cent of the nitrate could not be obtained in a sufficiently fluid state for pouring. Amatol became the principal high explosive for filling shells, mines, depth charges and bombs. To meet these requirements the production of ammonium nitrate was stepped up from nil in 1914 to 159,114 tonnes in 1916, a total of over 341,000 tonnes being made throughout the war.

The new explosives were not accepted, however, without a struggle. Lord Moulton was writing to Lloyd George, on 16 June 1915:

> Some two months ago the use of this mixture [Amatol] was approved and directions were given that it should be applied to the whole land service for six-inch [152 mm] shells downwards and experiments were directed to be made with regard to the larger shells. . . . To my regret I find that those who have charge of the loading of the shells have for the last two months completely disregarded the direction . . . It is hopeless for me to struggle to meet the extraordinary demands created by the War if there is on their part a disregard of or a reluctance to accept the necessary modifications of our artillery methods . . . and it is only by your coming to my help as Minister of Munitions that I can hope to obtain immediate and implicit acceptance of these all-important conditions of the supply to the Services.[10]

This hiccup was caused, mainly, by rivalry between generals and admirals. The Navy's refusal to accept Amatol for their shells made the Army think that it was being treated as a second-class service when it was directed, by a *lawyer*, to use something not thought to be suitable by the Navy. A few heads had to be knocked together and in the end it was just one of the many times throughout the war when it was necessary for people to be reminded of the truism that 'in spite of much evidence to the contrary, we are all fighting on the same side'.

*   *   *

Throughout the early months of the war all the shell-filling had been carried out at Woolwich but it soon became obvious that the establishment could not cope with the vast increase in the number of shell cases. On a visit to the Arsenal some months after the outbreak of the war, Lloyd George 'found stacks of empty shells which were being slowly and tediously filled, one at a time, with ladles from cauldrons of seething fluid'.[11] As soon as he could, he replaced the head of the Arsenal and began to set up national filling factories – one at Aintree and another at Coventry being opened in July 1915. Four more opened in August and six in September. By the end of the war there were eighteen. The number of shells filled in a week in September 1915 was 120,000. By January 1916 it had risen to 280,000, and by the middle of that year it was 1,180,000. In the year from July 1915 to June 1916, very nearly 20,000,000 shells were filled. And what was just as important, the number of heavy shells being filled grew most rapidly.

Much of the unpleasant and hazardous work was done by women and girls, and there were strict rules and regulations which had to be followed. Respirators, for example, had to be worn in some of the factories and the skin had to be protected from TNT dust by covering it with a special grease. There was a real chance of severe disfiguration, even death, from the jaundice caused by TNT poisoning. The first effect of this was to turn the skin a very bright yellow, which led to the workers in the filling factories being known as 'canaries'. Ninety-six of them died from the disease.

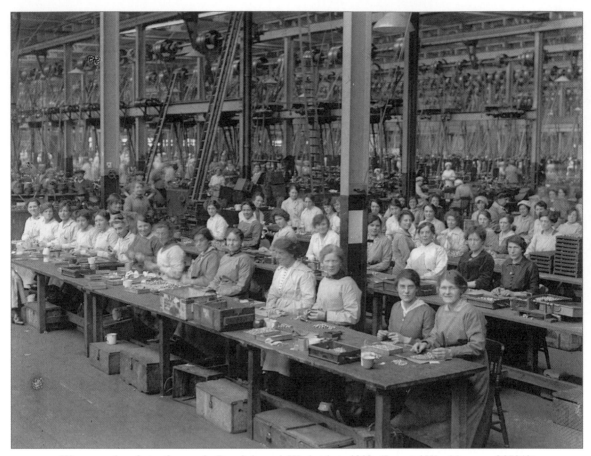

Women workers fitting fuses at the Royal Arsenal, Woolwich, in 1918. (Imperial War Museum, Q27862)

Even more deaths were caused by a single, horrendous accident, on Sunday 2 April 1916, in an Explosives Loading Co. Ltd factory on the site next to the Cotton Powder Co. in Faversham. Around noon on the fateful Sunday, a workman going for his lunch noticed that some empty bags, which had contained TNT, were on fire just outside a building in which 15 tonnes of TNT and 150 tonnes of ammonium nitrate were stored. The danger was at first underestimated, because it was considered that there was nothing present which could detonate the TNT. Alas, about an hour later it blew up, causing a number of other explosions in the area. Among the 108 dead and 64 injured were many who had been helping to put out the blaze, and even some who were simply spending their lunchbreak watching the scene. Sixty-nine of the dead were buried in a communal grave which is still a dignified feature of the Faversham Borough Cemetery. It is a measure of the full horror of an explosives accident that only about half of those interred could be positively identified.

The Faversham memorial. (Arthur Percival)

The filling factories were controlled by Lord Chetwynd, who built and supervised a large factory at Chilwell, which, within a year of its opening in January 1917 had filled almost 2,500,000 shells. He was an individualist of great ingenuity and courage. The cheapest form of Amatol, containing 80 per cent ammonium nitrate, could not be poured into shells through the nose as a liquid, which was the normal, relatively easy way of filling shells with pure TNT or the other Amatols. The research department at Woolwich suggested that this should be overcome by compressing the 80 per cent Amatol into cakes for inserting into shells through a detachable base. But this meant yet another design of casing and more delay, so Chetwynd developed his own method of pressing the Amatol in through the nose. The necessary grinding and apparent rough treatment of the explosive seemed to many to be highly dangerous. Chetwynd's response was to set himself up in a house very close to the plant, saying, 'If anyone is to be blown up, I'll be the first.'[12]

By the end of the war the Chilwell factory had filled 19,250,000 heavy shells, and the boldness of all concerned ensured that the British Army had more than adequate supplies of ammunition for the terrific bombardments that took place in the last two years of the war. In 1914 the British Expeditionary Force had only about 25 heavy guns and about 450 light ones. In 1916, the Somme offensive, over a

Checking shells at the Chilwell factory. (Imperial War Museum, Q30011)

30 km front, was supported by 450 heavy guns (one every 64 m) and over 1,400 light ones. In 1917, at the battle of Messines, on a 13 km front, 750 heavy guns (one every 18.3 m) and over 1,500 light artillery pieces were used; they fired 3,500,000 shells at an estimated cost of £17,500,000. By 1918, in an American assault on a 19 km front at Saint-Mihiel, over 3,000 guns were used (one every 6.4 m).

Explosives were also used increasingly in underground mining, one of the oldest arts of warfare. It was first practised in the East, possibly being used in the demolition of the walls of Jericho, and it was brought to Europe by the Crusaders. Immortalized by Shakespeare[13] in the lines:

> For 'tis the sport to have the enginer
> Hoist with his own petar; and't shall go hard
> But I will delve one yard below their mines,
> And blow them to the moon

the old art became an effective part of modern warfare once the opposing front lines were locked so close together for such long periods. At the start of the war the British Expeditionary Force had no facilities for underground mining but by the end of 1915 there were twenty-one Royal Engineer Tunnelling Companies containing many men who had been miners in civilian life. Their skills culminated in the most dramatic mining exploit of the war when many of the German positions on the top of the Messines Ridge were blown up at 3.10 a.m. on 7 June 1917. Some of the tunnelling had begun as early as August 1915 and by the time of the planned attack, which had been delayed many times, about 5.5 km of tunnels had been excavated and nineteen large loads of explosives had been placed under the enemy positions. Military Ammonal, a mixture of 65 per cent ammonium nitrate, 17 per cent aluminium, 15 per cent TNT and 3 per cent charcoal, was the main explosive used but it was augmented with guncotton and with Blastine, which contained ammonium perchlorate, sodium nitrate, dinitrotoluene and paraffin wax. Almost a million pounds of explosives was used in all, and the blast was heard in London and Dublin. One of the main mines, called 'Peckham', contained 30,000 kg of Ammonal, 6,800 kg of Blastine and 3,175 kg of guncotton. It completely obliterated a circle of ground with a diameter of 100 m. Of the blast itself, a British soldier who was at the scene wrote: 'The ground began to rock. My body was carried up and down as though by the waves of the sea. In front the earth opened and a large black mass mounted on pillars of fire to the sky, where it seemed to remain suspended for some seconds while the awful red glow lit up the surrounding desolation.'[14]

When Churchill replaced Lloyd George as Minister of Munitions in 1917 he was able to look forward imaginatively and hopefully. In his first munitions budget he wrote:

This is even more an explosives war than a steel war. Steel is the principal vehicle by which explosives are conveyed to the enemy. The requirements of propellant are limited by the production of shell, but there are other methods, besides those of artillery, of delivering high explosives to the enemy. The capacity of our existing plant is at present in excess of our shell programme for 1918. It has been arranged to supplement the present system of discharging high explosives by providing up to a maximum of 12,000 tons [12,192 tonnes] a week of bombs to be dropped by aeroplanes. The possibility of extending the Trench Mortar offensive is also being examined. To utilise fully our existing high explosive plants it is necessary that we should ship from Chile approximately 788,000 tons [800,608 tonnes] of nitrates; at present the tonnage for only 600,000 [609,600] has previously been agreed upon. New and very serious demands for TNT are also being made by the Admiralty for mines.[15]

The shell shortage and the scandal had been forgotten. But in the early days it had been a mighty close-run thing. Britain has always been good at brinkmanship; she has been saved, over and over again, from the disasters caused by doing too little too late by doing a lot as the last ditch came closer. And history was to repeat itself at the start of the Second World War in September 1939.

* * *

With the end of the First World War in 1918 the military role of TNT declined very sharply but more and more of it came to be used in explosive mixtures for mining and other commercial purposes. As war clouds gathered again in the 1930s the staff of the Director of Ordnance factories, no doubt remembering the chaos of 1914, began to revive the existing capacity for explosives manufacture. A patent for a new method of manufacturing TNT was taken out in 1932 and assigned to the Secretary of State for War, and a pilot plant was built at Waltham Abbey. This early planning meant that Britain was much better prepared, at least as far as explosives manufacture was concerned, in 1939 than in 1914, and during the Second World War twenty-four more plants, based on the earlier pilot plant, were erected. They produced more than 284,000 tonnes of TNT during the war.

The use of TNT in the war was also enhanced and supplemented by the introduction of two new explosives: PETN and RDX. Both substances had been known during the First World War but it had not been possible to make them on a large scale because of a shortage of methanol (wood alcohol). At that time the alcohol, obtained by heating wood, was both scarce and expensive and it did not become readily available until 1924, when the large German firm, I.G. Farbenindustrie A/G developed a new process at their synthetic ammonia plant at Leuna, using high-pressure techniques learned from operating the Haber process. The new process enabled methanol to be made, essentially, from coke and water.

PETN – pentaerythritoltetranitrate – was first discovered in Germany in 1894. It is a white solid, melting at 141°C, made from methanal (obtainable from methanol), ethanal and nitric acid. It is a very powerful explosive but too easily detonated and too sensitive to shock to be used on its own. It can, however, be desensitized by mixing it with waxes; it can be used as a component of detonators, fuses, primers and boosters; and it can be incorporated with plastic nitroglycerines or with synthetic rubbers to make a plastic explosive. It also enables better use to be made of TNT. In filling shells with molten TNT or Amatol it had always been difficult to avoid holes forming as the melt solidified but addition of 20–50 per cent of PETN prevented this; the resulting mixtures were known as pentolites. PETN, or Penthrite (or Penta, as it is known in some English-speaking countries), became almost as important as TNT and was used worldwide under such names as Pentrit, Niperyth, NP or Ten.

Cyclotrimethylenetrinitramine was first made by Henning in Germany in 1899 and was patented by him because he thought it might have medicinal uses. It is a white solid, melting at 204°C, made from methanal, ammonia and nitric acid. It was developed as an explosive by the research department at Woolwich between the wars, and they labelled it RDX (Research Department Explosive); it was later also known as Cyclonite or Hexogen. A pilot plant capable of making 255 tonnes of RDX was built at Waltham Abbey in 1938, and this was the sole source of supply until 1941 when other factories came into production.

RDX is as powerful as PETN but slightly less sensitive. It is used for filling shells, bombs and torpedoes either after being desensitized by mixing with waxes, or mixed with TNT. A mixture of 60 per cent RDX with 40 per cent TNT is called cyclotol; it develops a pressure of 270,000 atmospheres when it detonates.

Torpex, used for filling torpedoes, is a mixture of RDX, TNT and aluminium. RDX can also be mixed with about 20 per cent of jellies, waxes and plasticizers to make powerful, plastic explosives looking and feeling very much like plasticine, which are particularly easy to handle and to detonate. They are made in a range of slightly different compositions, to retain their plasticity over a span of temperatures, and they are ideal for demolition charges because the explosive can be pressed into any shape.

HMX – cyclotetramethylenetetranitramine – is closely related to RDX; it was first made in 1930 and has been used as an explosive since the 1950s. It is more powerful than RDX and is, currently, the most powerful explosive in service. A relative newcomer, HNIW – hexanitrohexazaisowurtzitane – is still more powerful and is undergoing tests. Looking further ahead, work is in progress to try to synthesize octanitrocubane and octa-azacubane, which should, theoretically, be very powerful explosives.

*　　*　　*

One particular form of plastic explosive, with the trade name Semtex, has been extremely well publicized in recent years because of its widespread use by terrorist organizations. The main component is PETN, plasticized by the incorporation of a styrene-butadiene copolymer. The explosive is made in large quantities by the firm Synthesia (formerly Explosia) situated at Pardubice, a small town 100 miles east of Prague, the capital of the Czech Republic. Large amounts of Semtex were supplied to the communist forces during the Vietnam War, but when that conflict ended in 1973 Czechoslovakia (as it then was) had a large surplus, much of which was sold to Libya. That country made it freely available to a number of terrorist organizations, including the IRA, thereby escalating what one commentator called 'the language of international violence'.

Semtex is particularly suitable for terrorist uses. It is very stable and stores well; it can be easily pressed into any shape because it is so malleable; it is very powerful so that it only requires about half a pound of the explosive, correctly placed, to down an aeroplane; and it is not readily detectable by X-rays. The original material was also odourless so that it could not be detected by sniffer dogs, but in 1991 the manufacturers responded to international pressure and incorporated a 'smell' which can be detected by airport security machines. Such machines have been extensively developed in recent years. The trail of terror spread by Semtex began around 1980, and, sadly, it seems that some terrorist groups still have enough of the explosive in their possession to perpetrate many more horrors.

*　　*　　*

To limit the damage that could be done by increasingly powerful explosives it became necessary, during the Second World War, to strengthen the armour-plating on tanks and ships, and to protect defence works by ever-increasing thicknesses of concrete. As a result, the war saw a fluctuating balance between the attacking power of one side and the defensive strength of the other.

Temporary supremacy could be gained by using more powerful explosives, but both sides also found it possible to make better use of their existing explosives by applying the principle discovered by Charles E. Munroe in 1888, when he was working at the Naval Torpedo Station at Newport in the United States, and which came to be known as the Munroe effect. He found that when a block of guncotton with letters carved into its surface was exploded against a steel plate, an impression of the letters was stamped into the plate. Similarly, if the letters on the guncotton were raised above its surface, they could be reproduced in relief on a metal plate. It was the explosive furthest away from the metal surface that had the most marked effect on the metal. Some time later, in 1910, a German, Egon Neumann, discovered independently that a block of TNT with a conical indentation in it would blow a hole through a steel plate that could only be dented by a similar weight of explosive without an indentation.

When an explosive is detonated, the shock wave moves at great speeds in all directions, like the light from a freely hanging bulb, but Munroe and Neumann had, surprisingly, found that the shock wave could be focused and concentrated mainly in one direction, in much the same way as the light from a torch bulb can be focused by the reflector.

The secret of making practical use of the effect lay in forming the explosive into what became known as a cavity or shaped charge. When a conical hole is made in one end of a cylindrical charge of solid pentolite or a mixture of RDX and TNT, and the charge is detonated at the other end, the shock wave is focused along the axis of the cylinder. If the conical hole is lined

Left: a Tallboy (Mk I M.C.) bomb. The bomb was 6.4 m in length and 1 m in diameter. It weighed 5,391 kg and contained 2,360 kg of explosive. (Crown copyright. The RAF Museum, Hendon) Right: the damage caused to the roof of a U-boat pen at Brest by a Tallboy bomb. (Imperial War Museum, CL1246)

with metal, a focused jet of extremely hot gases, moving at tremendous speed, is produced. This can strike a target so forcefully that it can drill a narrow, cylindrical hole through large thicknesses of concrete or several centimetres of armour-plating.

During the Second World War several designs of large, shaped charges were used for penetrating the concrete surroundings of pill-boxes or submarine pens which could resist ordinary missiles or bombs. The British PIAT (Projectile Infantry Anti-Tank), the American Bazooka, the Russian Simonov and the German Panzerfaust, which were all light weapons fired from across the shoulder, also enabled infantrymen to attack tanks with missiles containing a shaped charge. Modern anti-tank weapons include the Franco-German Milan and MR Trigat, the British LAW, the French Apila, the American TOW 2 and the Russian AT-4 Spigot. They generally fire either HEAT (high explosive anti-tank) or HESH (high explosive squash head) missiles. The former have shaped charges designed to penetrate a tank's armour. In the latter, the head of the charge is made of plastic explosive, and it is hoped that this will first spread out over the outer surface of a tank's armour-plating and then, on detonation, dislodge a lethal scab of armour from the inner surface.

A British LAW (Light Anti-Armour Weapon). This portable weapon weighs 10 kg and is 1 m long when carried but extends to 1.5 m when in use. It fires a 94 mm projectile with a shaped charge which can penetrate over 650 mm of tank armour at ranges up to 500 m. (Crown copyright: the Army Picture Library)

Tank designers have responded in three main ways: first, by using thicker and tougher steel armour; second, in the 1970s, by using Chobham armour, a laminated structure of steel, titanium amd ceramics, invented at Chobham in England; and third, in the 1980s, by using what is known as reactive armour. The basic armour-plate of a tank is covered, at its most vulnerable points, with a layer of explosive material in the form of 'tiles'. When hit by an incoming missile, the 'tiles' explode and damage the missile before it can reach the main armour of the tank. It is a very ingenious way of using one explosive to foil another. The counter-measure to reactive armour is to use a warhead containing one shaped charge behind another. The first will detonate the protective explosive layer on the tank, while the second, fired microseconds later, will penetrate the main armour.

Rival designers fight an on-going battle for supremacy between tank armour and anti-tank weapons and it will be interesting to see what the next step might be. TNT will probably not be involved in such considerations but it has certainly played a very large part in the history of explosives during this century.

CHAPTER 13
Setting It Off

In the early days gunpowder was always ignited by a flame or by a spark but during the nineteenth century chemicals which were so sensitive that they could be exploded by a sharp blow or by friction as well as by heat, came to be used. Potassium chlorate, the sensitivity of which had been discovered when Berthollet tried to incorporate it into a gunpowder mixture in 1788, was a common choice, along with a group of compounds known as fulminates. They were the first examples of what are nowadays called primary or initiating explosives. Although they are particularly sensitive they are not required in very large quantities so that special precautions can be taken in their manufacture and handling.

Fulminating gold or aurum fulminans (from the Latin *fulmen* meaning lightning) is an olive-green solid of uncertain composition made from gold oxide and ammonia. It was described by Basil Valentine in 1603 as 'a powder which kindles as soon as it takes up very little heat or warmth, and does remarkably great damage when it explodes with such vehemence and might that no man would be able to restrain it'.[1] And it is referred to in Samuel Pepys's diary for 11 November 1663, when the entry reads:

> At noon to the Coffee-house, where, with Dr. Allen, some good discourse about physic and chemistry. And among other things I telling him what Dribble, the German doctor, do offer of an instrument to sink ships: he tells me that which is more strange that something made of gold, which they call in chemistry Aurum fulminans, a grain, I think he said, of it put into a silver spoon and fired, will give a blow like a musket, and strike a hole through the silver spoon downwards, without the least force upward.[2]

Dr Allen was an old Cambridge friend of Pepys while 'Dribble, the German doctor', seems to refer to Jacob Drebbel, the son of Cornelis Drebbel, a well-known Dutch inventor famous for his perpetual motion machine, for his submarine (capable, it was said, of travelling for several hours under water at a depth of 3–4 metres), and for his machine that could produce rain, thunder, lightning or cold to order. He spent long periods in England, where he was favourably received and patronized by James I, and he served in the British Navy for a time around 1620. It was in that capacity that he claimed to have found a way of using aurum fulminans in making mines which could sink ships and which were tried out in an expedition against La Rochelle in 1628.

Some years later Dr Kuffler, Jacob Drebbel's brother-in-law, approached Pepys, who at the time was Clerk of the Acts to the Navy Board, with a suggestion that the mines should be adopted by the Navy. His diary entry for 14 March 1662 records the encounter:

> In the afternoon came the German, Dr. Kuffler, to discourse with us about his engine to blow up ships. We doubted not the matter of fact, it being tried in Cromwell's time, but the safety of carrying them in ships; but he doth tell us that when he comes to tell the King his secret (for none but the Kings successionally, and their heires, must know it), it will appear to be of no danger at all. We concluded nothing, but shall discourse with the Duke of Yorke tomorrow about it.[3]

It is claimed that Jacob Drebbel and Dr Kuffler demanded £10,000 if the invention was found to be successful so they must have thought highly of it, but nothing emerged and fulminating gold is too sensitive to handle at all safely in anything but very small quantities, and it is also very expensive. It was really only suitable for use in fireworks and toys and it was replaced in that use by silver fulminate, a black powder prepared by Berthollet in 1788. This was very similar to fulminating gold, but much cheaper, and it was used as early as 1803 in demonstrations of supposed magic by charlatans and mountebanks at markets and fairs. Christopher Grotz gave details of how to make 'amusements with fulminating silver' in his book *The Art of Making Fireworks, Detonating Balls, etc.*, published in 1818. A joke cigar, for example, is made by 'opening the smoking end, and inserting a little of the silver; close it carefully up, and it is done'. A joke spider 'by cutting a piece of cork into the shape of the body of a spider, and a bit of thin wire for legs. Put a small quantity of the silver fulminate underneath it; and on any female espying it, she will naturally tread on it, to crush it, when it will make a loud report.'

More important applications of fulminates for military and mining purposes had to await the detailed report[4] given to the Royal Society on 13 March 1800 by Edward Howard on 'a new fulminating mercury'. He was an amateur, gentleman scientist, the brother of the 12th Duke of Norfolk, and a man of some courage. He was severely wounded in an explosion which destroyed most of his apparatus and led him to write: 'I must confess I feel more disposed to prosecute other chemical subjects.'[5] But he was not deterred and his research had a very significant effect on the history of explosives, first in the development by the Revd Alexander Forsyth of a new method of firing a shotgun, and later in military uses and in mining.

*   *   *

The Revd Alexander John Forsyth was born on 28 December 1769 in the manse at Belhelvie in Aberdeenshire where his father was the minister. After graduating at King's College, Aberdeen, and being licensed as a preacher, he succeeded to the living in Belhelvie at the age of twenty-one, after his father's sudden death. There were few signs, at that stage, that the country cleric would make a name for himself in the world of explosives but he had become interested in mechanical inventions and chemistry at college and he was a very keen sportsman. In the time he could spare from his ministry he spent hours on the shores of a loch, close by the manse, shooting (or trying to shoot) wildfowl with a long 12-bore flintlock shotgun, but he was frequently exasperated when he found that the flash from his flint gave the birds sufficient warning to escape his carefully aimed, but somewhat delayed, shot. Fortunately one hobby came to the help of another for he had a workshop in a shed in the manse garden, known to the parishioners as the 'Minister's Smiddy', in which he designed a revolutionary new lock for sporting guns which overcame the problem and quickly replaced the flintlock for sporting purposes.

His first idea had been to try to shield the flash from the flintlock by constructing a hood to fit over the barrel of the gun, but this had not been very satisfactory so he turned his attention to chemical mixtures which could be exploded by a sharp blow or percussion. As he phrased it in his eventual patent,[6] 'instead of giving fire to the charge in the gun by a lighted match, or by flint and steel, or by any other matter in the state of actual combustion . . . I do make use of some or one of those chemical compounds which are so easily inflammable as to be capable of taking fire and exploding without any actual fire being applied thereto, and merely by a blow . . .'. He experimented first with fulminating mercury but found that it exploded too violently for his purpose so that most of his early experiments were done with mixtures of potassium chlorate, charcoal and sulphur.

What came to be known as the percussion lock was operated by a pull on the trigger causing a hammer to fall on to such a mixture positioned next to the touch-hole or vent in a gun. The success of this new method, in tests carried out on the Scottish loch during the winter of 1805/6, was bad news for wildfowl but good news for sportsmen, and some of those who were in the know encouraged Forsyth to exploit his invention and to investigate its use for military purposes. To that end, he took his idea to London in the spring of 1806, where Lord Moira, the Master General of Ordnance, was sufficiently impressed to obtain leave of absence for him from his parish duties, to arrange for him to carry out further work in the Tower of London and to pay him an advance of £100. There were many technical problems to be solved because, although an occasional misfire does not matter much when shooting wildfowl, consistency is essential for military purposes. By April 1807, however, after much patient and dangerous work, most of which he had had to do by himself because others did not like handling the dangerous materials involved, Forsyth submitted a carbine and a 3-pounder gun operated by his percussion lock. Both were approved by Lord Moira, who suggested that a payment should be made equal to the value of the gunpowder that would be saved by using the new lock. Alas, at that important juncture there was a change of government and Lord Moira was replaced as Master General of Ordnance by Lord Chatham, who took a very different view. He told Forsyth to stop all his work, to return any government property he might have, to remove all his own 'rubbish' from the Tower, and to claim his expenses. They came to £603 18s 4d, and they were paid in 1807.

Forsyth was disappointed but not defeated for he knew that his idea was a good one. So, too, did Napoleon, and there is a traditional story that he offered Forsyth £20,000 for his invention but that this offer was flatly rejected. Realizing that the military sphere was closed to him, at least for the time being, Forsyth directed all his efforts into the market for sporting guns. With the help of James Watt, he took out a patent in July 1807, set up his own company and opened a gun shop at No. 10 Piccadilly, employing James Purdey, the later founder of the famous firm of James Purdey & Sons, as an assistant. The locks made by Forsyth & Company, Patent Gun Makers, were based on the design Forsyth had perfected in the Tower of London but they were beautifully decorated in accordance with the practice of the day. In place of the flash-pan in a flintlock gun, a steel plug, with a 3 mm hole down the centre was screwed through the barrel giving access to the charge of gunpowder in the chamber of the gun. Above the plug there was a small flash-pan on to which a little detonating primer mixture could be dropped from a rotatable magazine, shaped like a scent bottle and containing enough primer for about twenty shots. When the magazine was rotated back into its original position it moved a striker-pin over the top of the primer. Pulling the trigger released a hammer which fell on to the pin and exploded the primer. Advertisements claimed:

> The Forsyth patent gunlock is entirely different from the common gunlock. It produces inflammation by means of percussion and supersedes the use of flints. Its principal advantages are the following: The rapid and complete inflammation of the whole charge of gunpowder in the chamber of the barrel. The prevention of loss of force through the touch-hole. Perfect security against rain or damp in the priming. No flash from the pan and less risk of accidental discharge of the piece than when the common lock is used. The charge of gunpowder to be from one third to one fourth less than when the flintlock is used.[7]

The claims proved to be well founded with over 4,000 locks being sold, even though they were rather expensive, between 1808 and 1821 when the patent expired. But Forsyth had not become rich because he had been involved in endless litigation to protect his patent rights. With Lord Brougham, to whom he was related, as his trusted counsel he successfully fought one case in 1811, one in 1816, one in 1818 and three in 1819. By then, percussion locks were commonplace among sportsmen and the legal cases were a measure of how keen other gunsmiths were to get in on the act, but it was not until 1838 that the system was adopted by the Army.

At that stage Forsyth, who had sold his share in the gun business and returned, somewhat embittered, to the manse, was persuaded by Lord Brougham and other friends to claim compensation and, following a petition to Parliament, the Lords of the Treasury granted him 'a gratuity of £200 for remuneration as the original inventor of percussions for small arms' in 1842. But Forsyth was found dead at his breakfast table on 11 June 1843, before he had learned of this award, and he was buried in Belhelvie churchyard alongside his father and among the

PER ASPERA TENAX

1768   1843

To the Memory of the Reverend
ALEXANDER·JOHN·FORSYTH
M·A. LL·D
Minister of Belhelvie
Aberdeenshire
In 1805 he conducted experiments
in the Tower under the Master
General of Ordnance and in 1807
invented the percussion system
which was adopted by the
British Army in 1839
This Tablet was erected
in 1929 by admirers of his
Genius

The Forsyth Memorial Plaque.
(The Board of Trustees of the Royal
Armouries)

parishioners whom he had served so faithfully for fifty-two years (when he had not been designing or shooting guns). There must have been some guilty consciences about the tardiness and the paucity of Forsyth's payment because, after his death, his niece was paid £500 and his grand-nephew and sister's husband each received £250. A Forsyth medal was struck, to be awarded to the winner of a shooting competition at Bisley, and in 1929 a plaque was erected to his memory in the Tower of London.

* * *

Forsyth's idea was so good, so practical and so marketable that many other gunsmiths tried to emulate it and some improvements had been suggested and tried even before the patent rights ran out. In particular, Pauly, a Swiss gun-maker working in Paris, and Joseph Manton in England made significant advances which signposted the way forward. In 1812 Pauly added a little gum arabic to a mixture of potassium chlorate, charcoal and sulphur and compressed it into a pellet between discs of paper, which was very much like the present-day cap used in toy guns. Two years later Manton not only used similar pellets but also designed a tube-lock in which fulminating mercury was contained in a copper tube. Both had the advantage of doing away with finely powdered percussion mixtures, which were not easy to handle, and they were the forerunners of modern percussion caps.

Originally, these caps were shaped like a very small top-hat and they were made first of steel, then of pewter and finally of copper. They contained a small amount of percussion powder or primer covered by tin foil, the whole unit being sealed and waterproofed by a coating of shellac. The cap was placed with the tin foil in contact with a thin metal tube, or nipple, within the gun and leading on to the gunpowder in the chamber. A blow of the hammer on the cap exploded the primer and broke the foil so that the flash was transmitted through the nipple on to the gunpowder.

The invention of the cap was claimed by almost all the contemporary gunsmiths but the balance of the evidence favours an English landscape artist, Joshua Shaw. He was an even less likely candidate than Forsyth to join the hall of fame in the realm of explosives for he was not even a mechanic and he had to get his designs fabricated by someone else. But he recognized the potential of his idea and as he was prevented from developing it in Great Britain by Forsyth's patent rights he emigrated to the United States in 1816 and settled in Philadelphia. He was not at first much better off because he could not get a patent there as he was not an American citizen at that time, but he did begin to work for the government on making percussion caps for military weapons. Like Forsyth, he discovered that governments are not prompt payers and a claim for compensation which he made in 1831, after an explosion in which his left hand was disabled, was only met in 1847 when he was awarded $16,000 for all his patent rights.

Meanwhile, in England Joseph Manton, encouraged by his patron, Colonel Peter Hawker, began to sell arms with copper-cap percussion locks as soon as Forsyth's patent expired, and Joseph Egg, a London gunsmith, E. Goode Wright of Hereford, and the firm of Messrs Joyce & Co. were also making similar guns around the same time.

Thereafter, there were continual changes in the composition of the percussion mixture used in caps, and when breech-loading guns were introduced in the second half of the nineteenth century the cap, the gunpowder and the projectile were assembled together in a single cartridge very much like those used in today's shotguns and rifles. In some larger guns the propellant, with a cap to fire it, was packaged separately from the projectile.

Forsyth had discovered that mercury fulminate exploded too violently to be used satisfactorily on its own as a percussion powder because it caused serious deformation when used in a cap. Its action therefore had to be moderated and this was originally done by mixing it with gunpowder, or with potassium chlorate and either powdered glass or antimony sulphide, and these mixtures were used for many years until gunpowder was replaced as the propellant by cordite. It was then necessary to use a new initiating mixture because it was more difficult to detonate cordite than gunpowder and a typical cap composition contained 19 per cent mercury fulminate, 33 per cent potassium chlorate, 43 per cent antimony sulphide and 2.5 per cent each of sulphur and gunpowder. Similar mixtures were in vogue for many years.

*   *   *

Mercury fulminate was introduced into mining at the same time as Nobel discovered nitroglycerine in 1864. At first he detonated the nitroglycerine by using a small charge of gunpowder contained in a glass bulb or a hollow wooden cylinder. Later he used a mixture of gunpowder and mercury fulminate, or the fulminate by itself, packed into a copper cylinder closed at one end. He referred to this as his patent detonator, and that term is still used in military parlance, but the device is generally called a blasting cap, or just a cap, in mining circles.

The cylinder of a modern detonator is generally made of aluminium and there are various, numbered sizes, with the commonest, the No. 6, being about 40 mm long and 6 mm in diameter. The closed end is slightly concave so that it produces a slight cavity effect which improves its performance. The detonator is filled about one-third full with the initiating mixture and, in a plain detonator, it is fired by inserting a length of fuse into the open end and crimping it in position. Alternatively, in an electric detonator, firing is achieved by passing a low voltage current through a fine wire embedded in a flashing composition in a fuse head sealed into the top of the detonator. The current heats up the wire which then ignites the flashing composition which, in turn, sets off the initiating mixture. In modern detonators it has become technically possible to incorporate a delay element so that it fires after a specific time and not instantaneously. Detonators with different delay times, varying by approximately 25 milliseconds, are available.

Mercury fulminate, either alone or mixed with potassium chlorate, potassium nitrate or gunpowder, was used for filling detonators for many years, but it was not entirely satisfactory because it was very expensive and did not store well in hot climates. To limit the amount of fulminate that was needed, compound or

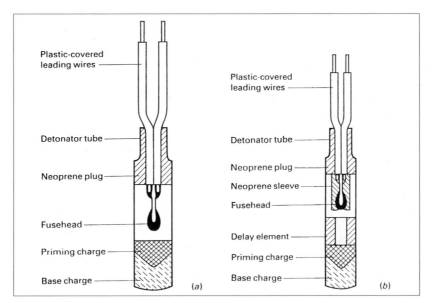

Electric detonators: (a) instantaneous; (b) with an inbuilt delay. (Chemistry in Britain)

composite detonators were introduced at the start of the twentieth century, Lyddite or TNT being compressed into the lower part of the detonator as what was called the base charge, with only a small amount of the initiating explosive on top of it, perhaps covered by a perforated metal disc.

About the same time, too, mercury fulminate was beginning to be replaced by lead azide. This is a white solid, first made by Theodor Curtius in Germany in 1891, and it is both cheaper and stores better than the fulminate. The first attempts to use it were hindered by some fatal accidents, but patents for its use in detonators were taken out in Germany and France in 1907, and it was in military use during the First World War. It was not, however, until the 1930s that lead azide had almost completely replaced mercury fulminate in mining detonators.

It is a more powerful initiating agent than the fulminate but it is not entirely reliable when used on its own because it is not always ignited by the heat from the fuse. So in a modern detonator it is mixed with 25–50 per cent lead styphnate which is more susceptible to heat. The base charge is either PETN, tetryl or RDX, with the first-named being by far the commonest. Diazodinitrophenol (DDNP or DINOL) and tetrazene are also used as initiating explosives in detonators instead of lead azide.

* * *

Plain detonators go, almost literally, hand in glove with the special safety fuse invented by William Bickford in 1831. A philanthropic Methodist, born in Devon in 1774, his early life was both uneventful, and, from a business point of view, not very successful. He was a leather merchant, first in Devon, then in Truro in Cornwall, and latterly in Tuckingmill, near Cambourne, the centre of the Cornish mining industry. Bickford soon became so distressed by all the local accidents, associated with handling explosives in mines, that he turned his mind to providing a solution even though he had no scientific knowledge and no great pretensions to learning. By chance, he saw a friend spinning some yarns into a rope by pulling and twisting them as he walked backwards, and with a flash of genius it came to him that he might be able to feed a trail of gunpowder into the centre of the rope as it was being spun.

With the assistance of his daughter's husband, George Smith, who was a carpenter and builder, and Thomas Davey, a working miner – both fellow Methodists – he found that this simple idea could be made to work, and a patent, No. 6159, was taken out on 6 September 1831, when Bickford was fifty-seven years old. Beginning with the phrase 'To all to whom these presents shall come, I, William Bickford, of Tuckingmill, in the County of Cornwall, Leather Seller, send greeting', the patent continued:

> I embrace in the centre of my fuse, in a continuous line throughout its whole length, a small portion or compressed cylinder, or rod of gunpowder, or other combustible matter prepared in the usual pyrotechnical manner of fire-work for the discharging of ordnance, and which fuse so prepared I afterwards more effectually secure and defend by a covering of strong twine made of similar

William Bickford. (ICI)

material, and wound thereon at nearly right-angles to the former twist, by the operation which I call countering, hereinafter described, and I then immerse them in a bath of heated varnish, and add to them afterwards a coat of whiting, bran, or other suitably powdery substance to prevent them from sticking together or to the fingers of those who handle them; and I thereby also defend them from wet or moisture or other deterioration, and I cut off the same fuse in such lengths as occasion may require for use. Each of these lengths constituting, when so cut off, a fuse for blasting of rocks and mining, and I use them either under water or on land, in quarries of stone and mines for detaching portions of rocks, or stone, or mine, as occasions require, in the manner long practised by and well known to miners and blasters of rocks.

William Bickford did not live to see the full success of his invention for he became seriously ill a year after his original patent had been taken out, and he died in 1834, but the business he had started grew rapidly under the control of his son (originally a schoolmaster), his daughter and her husband, and Thomas Davey. International acceptance of the fuse was helped, too, when it was adopted by the British services in the 1840s. It had been demonstrated to Colonel Pasley and Colonel Burgoyne at the War Office and the former had chided himself for having been too stupid to miss the opportunity. 'Here have I', he said, 'with the Arsenal behind me been all these years trying to scheme a safe and simple means of conveying fire to the blasting charge, and never thought of trying to make black gunpowder burn slowly and regularly, which a Cornishman has discovered in a rope walk.'[8]

The first factory was built at Tuckingmill in 1831, and 72 km of fuse, valued at £500, was made in the first year. A hundred years later, the factory made almost 170,000 km valued at £460,000, and it remained in operation until 1961 when manufacture of the fuse was transferred to Ardeer in Scotland. Plants were opened in America in 1836, in France in 1839 and in Germany in 1844, and by the end of the century the fuse was also being made in Spain, Austria, Australia and Hungary.

Once the original patent ran out in 1845, the firm of Bickford Smith & Company, as it eventually came to be known, had to contend with a great deal of competition and could only hold its own by constant improvement and innovation to meet new requirements. In particular, the waterproofing and strength of the fuse was improved by coating it with gutta–percha, a rubber-like gum from a Malaysian tree, and by various tapes and varnishes; the rate of burning was controlled more accurately; special colliery fuse, which emitted no sparks, was made; and various patent lighters for igniting the fuse were invented. Bickford Smith & Company remained a family firm for many years and the family must rank high on any list of benefactors to all users of explosives.

*   *   *

The firm also contributed greatly to the development and manufacture of instantaneous and detonating fuses. These burn or explode very rapidly and they are used to connect up a number of separate explosive charges so that they can all be set off at once. It is particularly useful to be able to do this in many mining, quarrying and tunnelling applications when it is necessary to remove large quantities of rock or earth in one operation, but there are also many military uses.

If the gunpowder trail in safety fuse is packed too loosely the build–up of pressure and the free flow of hot gases within the core of the fuse, when it is ignited, causes it to burn too rapidly. If this happened accidentally and unexpectedly, as in a so–called 'running fuse', the results could be disastrous, so great care had to be taken to ensure that the powder in safety fuse was well and evenly packed. It was, however, possible to make use of the phenomenon of rapid burning as in the first type of instantaneous fuse.

Strands of cotton were soaked in potassium nitrate solution and, before they were quite dry, they were coated with a layer of gunpowder by passing them through a suspension of finely divided powder and gum in water or methylated spirits. The strands, after thorough drying, made what was called quick match. When unenclosed, so that the gases could escape freely without any build–up of pressure, they burnt at a rate of about 76 mm per second, but if loosely enclosed in a strong, waterproof wrapping the rate of burning went up enormously. This was the basis of an instantaneous fuse manufactured in France as early as 1855, and in Cornwall in the 1880s. It went off with a loud bang but the rate of

A line of drying wheels used in the modern manufacture of safety fuse. (Crown copyright: Royal Commission on the Ancient and Historical Monuments of Scotland)

propagation was very irregular, varying between 30 and 150 m per second, and it was necessary to use special T- or X-shaped connecting links to join lengths of the fuse together when branch lines were required, and each separate explosive charge required its own detonator.

The fuse was nevertheless used very widely for many years until it was replaced in around 1907 by another French invention, known as *cordeau détonant* ('detonating cord'), or simply *cordeau*. This was made by pouring hot, liquid TNT into a lead tube with an external diameter of about 20 mm and a 13 mm bore and allowing it to cool and solidify. The filled tube was then drawn through a series of dies to reduce the diameter to about 6 mm and pulverise the TNT into a very fine powder. This fuse had a very regular rate of propagation around 4,900 m per second. It also had the further advantage over the earlier instantaneous fuse of being able to detonate all explosives simply by passing it through them and that did away with the use of a lot of separate detonators. The new fuse was authorized for use in Great Britain in 1909 and Bickford Smith & Company started to manufacture it in 1911.

Cordeau détonant was used well into the 1930s but it then largely gave way to an improved version, called Cordtex in Great Britain and Primacord in the United States, consisting of a thin trail of PETN within a textile and plastic coating. It

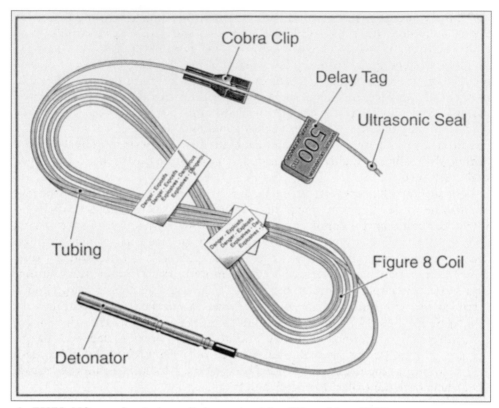

An EXEL MS non-electric (non-el) detonator with a 500 ms delay. (ICI)

had a rate of propagation of around 6,400 m per second and went off with a tremendous bang. It was cheaper than *cordeau détonant* and it was so light and flexible that it could be tied together with ordinary knots.

Both safety fuse and detonating fuse have been replaced to some extent by a newer non-electric fuse called Non-El, which was developed in Sweden by Nitro Nobel AB. This is made by spraying the inside of a thick-walled plastic tube with a thin layer of high explosive, and when one end is ignited by a blank cartridge there is a flash through the tube at a rate of about 2,000 m per second which can set off detonators embedded in explosive charges at the other end. The fuse on its own can replace detonating fuse and, in conjunction with delay detonators, it can replace safety fuse. It also has the advantage over electric detonation of being immune to any stray electric currents.

*   *   *

The precise moment at which an explosive is initiated may be of paramount importance. It is, for instance, no use firing at a bird after it has flown away. An individual can apply a match to a definite length of safety fuse, or pull a trigger on a gun, or remove the safety pin from a grenade and throw it, or switch on an electric current to fire an electric detonator, but such methods do not meet all requirements and great ingenuity has been applied in making all sorts of devices for setting off an explosion in the most advantageous way at precisely the right moment. Clockwork mechanisms, as in an alarm clock, a cooking timer or a parking meter clock, are very readily available and can be adapted without too much difficulty to release a striker against a percussion cap or to close an electric circuit at a pre-set time. They range from amateur, home-made efforts, which commonly misfire, to highly sophisticated, very accurate timers which can be set well in advance.

Less accurate, but cheaper, time switches, which rely on a chemical reaction, can also be made. The time-pencil used by saboteurs during the Second World War is the best known example. Over twelve million of them were manufactured during the war. They were shaped like a 15 cm long pencil and contained a wire inside a thin copper cylinder at one end, which held back a spring-loaded striker pin. The cylinder also contained a glass ampoule full of an acidic solution. Squeezing the copper cylinder broke the ampoule, and the solution slowly corroded the wire until it snapped and released the striker. The time taken for this to happen could be varied by altering the thickness of the wire or the concentration of the acidic solution, and six pencils were issued with delays between 10 minutes and a month, each with its own colour code. A similar time-switch, known as the L-delay, depended on the time it took for a thin rod of lead, which held back the striker, to stretch and break.

Neither was very accurate in its timekeeping and both were severely affected by changes in temperature. In cold weather they both took much longer. It may well have been the failure of a time-pencil to operate quickly enough that ruined an attempt by von Tresckow to assassinate Hitler in March 1943. An explosive charge fired by a ½-hour time-pencil and disguised as a bottle of Cointreau had been placed aboard his aircraft as a supposed gift for one of his staff. When the bomb did not go off, well after the allotted time, it had to be recovered by the disappointed conspirators on the somewhat flimsy pretext that there had been a bit of a muddle. Benjamin Disraeli may have thought, after the death of Abraham Lincoln, that 'assassination has never changed the course of history', but it came close to doing it in March 1943.

Another group of switches depend on a pull or a push or a change in pressure releasing a striker or closing an electric circuit. Simple examples are provided by the pull-switch operated by a trip-wire in a booby trap, by the railway fog signal which is placed on the line and exploded by a train running over it, and by anti-personnel mines which go off when they are trodden on. Similar devices, disguised as stones or animal droppings, were also available for use by saboteurs during the Second World War.

That war also saw many more sophisticated examples of the same basic idea. A pressure switch, placed under a railway line, was operated by the slight depression of the rail when the train passed, and could even be set to explode the attached charge when either the first, second, third or subsequent train passed.

There was also an altimeter switch for exploding a charge in an aeroplane only once it was airborne; it operated when the plane reached a certain altitude and the pressure fell accordingly. There was a switch attached to a magnetic, limpet mine, for placing on the outside of a ship, which would only operate once the ship was at sea and had reached a certain speed. Depth charges for attacking submarines which blew up when they reached a certain depth and the hydrostatic pressure rose sufficiently. The 'bouncing bomb', containing 3,000 kg of RDX, which destroyed the Möhne dam, was operated by three hydrostatic pistons set to fire when the bomb was 9 m under the water. Secret mines, kept in reserve by the Germans until just after the Allied invasion of Europe had begun in June 1944, contained a rubber bag and diaphragm designed to respond to a 0.1 per cent change in hydrostatic pressure, which could be caused by the passage of a ship overhead.

And there were, and are, many other possibilities, too, such as a small switch attached to a charge, which operated after immersion in petrol, for dropping into the tanks of German vehicles. Magnetic mines and acoustic mines operate respectively by the magnetic field or engine noise of a passing ship. Inertia switches commonly contain a globule of mercury which moves to make an electric contact when the switch is displaced, as, for instance, when a parked car is driven away. Photoelectric switches operate when a train, for example, goes into a tunnel. And, more recently, switches operated by infra-red light or by radio waves have come on the scene. Many of these timing devices have been ingeniously adapted by terrorist organizations, all over the world, to help them to perpetrate their evil deeds.

Such a high degree of technical sophistication is so commonplace nowadays, and the professionals have taken over so much from the amateurs, that it is difficult to remember that it was all started by a French Count, a Scottish clergyman, and two Englishmen, one a gentleman scientist and the other an artist.

CHAPTER 14

Nuclear Fission

Tinian is a long, thin island, approximately 16 km by 3, at the southern end of the Mariana group about 2,300 km from Tokyo. It was captured from the Japanese by the Americans on 1 August 1944, and rapidly transformed into the largest air base in the world with six 2 mile long runways. The island reverberated from the impact of hundreds of B-29 Superfortresses, which carried out routine, conventional bombing and incendiary attacks on Japanese cities, but a very special mission was launched on 6 August 1945.

At 0245 hours local time a heavily loaded, newly modified B-29 took off from the base, piloted by Colonel Paul Tibbetts, the commander of the 509th Composite Group. He had christened the plane *Enola Gay*, his mother's names, and the words were painted on the fuselage. Exactly 7 hours, 30 minutes and 30 seconds later the plane dropped the first ever atomic bomb, from a height of 9,631 m, on to the city of Hiroshima. The bomb was code-named 'Little Boy'. It consisted of a modified gun barrel, sealed at both ends and contained in a cylindrical outer casing. With fins to give it stability when dropped, it was about 3 m long and 0.75 m in diameter. It only weighed about 4,100 kg but it had the same power as 12,700 tonnes of TNT, so that it completely dwarfed any previous single explosion.

Conventional bombs were generally fused to explode when they hit the ground so that much of their blast effect was directed upwards into the air, which limited the damage they could do. In contrast, the atom bomb was set off at a height of 580 m so that its particularly powerful blast, much of it directed downwards, could cause damage over a large area of the ground below. The blast destroyed 60 per cent of the city and flattened an area of 9 sq. km. of buildings, many of which had been specially constructed to withstand the effects of earthquakes. By the time the bomb went off, Colonel Tibbetts had had about 45 seconds to take evasive action and his plane was more than 16 km away but he and his crew still felt the effects of the blast.

But the blast effect was not the only, nor the worst, aspect of the bomb. There was a tremendous flash of light and heat as the bomb went off and it was later estimated that the temperature at the point of explosion reached almost 3,000°C. People immediately below the bomb were scorched into small piles of black ash; many of those who were out in the open as far as 2½ miles away were severely burnt; telegraph poles at the same distance were charred or set on fire; the surface of granite, 600 m away, was partially melted; and much of the city was aflame for days.

Those who were not blown or burnt to death thought that they had survived but many of them, within a short time, began to realize that they were not feeling at all well, and it slowly began to dawn that they were suffering from radiation sickness caused by exposure to an overdose of gamma-radiation from the exploding bomb. The main, immediate effect was to disrupt the regeneration of blood cells within the body which led to bleeding, infection or anaemia. Some lucky ones died quickly. Others lingered on, suffering from the insidious, incurable symptoms for many years, before their turn came.

Because of the on-going deaths, it is difficult to quantify the calamity accurately. There were about 350,000 people in Hiroshima at the moment the bomb went off. Shortly afterwards the Japanese authorities estimated that 71,000 had been killed and 68,000 injured, but, over the years, the mortality figures grew remorselessly until more than 50 per cent of those who had been there on 6 August were thought to be dead.

As the Japanese government did not surrender, a second plane, named *Bock's Car* after its usual pilot's name, but flown on this mission by Major Charles W. Sweeney, took off on 9 August with the intention of dropping a second bomb on Kokura. That city, however, was covered by cloud so, after three abortive attempts to find the target, Major Sweeney flew on to his secondary target, which was Nagasaki. Visibility was still not very good when he arrived there but, at the last moment, and when he was beginning to run short on fuel, a hole opened up in the clouds and the bomb was dropped. It was 1102 hours. This bomb, code-named 'Fat Man', was of a different type from 'Little Boy'. It was egg-shaped, about 3.5 m long with a maximum diameter of 1.5 m and it weighed about 4,536 kg. 'Fat Man', equivalent to 22,350 tonnes of TNT, was more powerful than 'Little Boy' but it caused less damage because it was dropped slightly off target and because the terrain of Nagasaki was hilly. Nevertheless, 45 per cent of the city was destroyed and the final death toll was thought to exceed 50 per cent, as at Hiroshima.

The Japanese offered to surrender on 10 August, without knowing that there were no other atomic bombs available for immediate use.

* * *

Both 'Little Boy' and 'Fat Man' operated on the same basic principle. One atom of an element within the bombs was split by being bombarded with a neutron. This released a small amount of energy but, at the same time, produced more neutrons. These so-called secondary neutrons were then able to split more atoms so that a chain reaction was set up and as this happened very rapidly large amounts of energy were released. There was, however, an important difference between the two bombs: 'Little Boy' contained uranium-235 whereas 'Fat Man' contained plutonium. Both these elements were obtained from naturally occurring uranium, a very dense silver-white metal, which had been found in an ore called pitchblende by a German chemist, Martin Heinrich Klaproth, in 1789. He named it after the planet Uranus which had been discovered eight years earlier.

The explosion of the atomic bomb dropped on Nagasaki. The smoke is billowing out 6 km above the city. (Imperial War Museum, MH 2629)

The damage caused at Hiroshima. (Imperial War Museum, MH29447)

For many years uranium, in common with other elements, was regarded as being made up of one particular kind of atom, which, according to the Atomic Theory propounded by John Dalton in 1808, was indivisible and indestructible. But in 1919 F.W. Aston demonstrated that it actually contained three kinds of atoms, which were called isotopes. The three isotopes are chemically alike but they differ in the mass of their atoms. It is like having a family with identical triplets differing only in their weights. The isotopes are called uranium-238, uranium-235 and uranium-234 and they are symbolized as ^{238}U, ^{235}U and ^{234}U. The mass of the ^{238}U atom is 238 times that of a single atom of hydrogen; that of ^{235}U is 235 times heavier than the atom of hydrogen; and so on. Natural uranium contains 140 atoms of ^{238}U for every 1 atom of ^{235}U, and minute traces of ^{234}U.

The ^{235}U used in 'Little Boy' had to be extracted from naturally occurring uranium, and the plutonium in 'Fat Man' had to be made from ^{238}U because it does not occur naturally. There were, then, formidable problems in obtaining the raw materials from which to make the bombs, besides the difficulties of designing them successfully. The whole story is one of quite remarkable technical achievement.

*   *   *

It was the discovery of radioactivity that led to the first vague hints that it might one day become possible to release large amounts of energy from atoms. Henri Becquerel was investigating various phosphorescent substances in 1896, when he found that uranium compounds emitted a penetrating radiation capable of passing through the supposedly impervious wrapping of a photographic plate. On 16 and 17 February he sprinkled some uranyl potassium sulphate on to the black paper wrapper of a flat, unopened photographic plate and put it in a closed, dark drawer. On unwrapping and developing the plate, after two or three days, he found, much to his surprise, a clear image showing a silhouette of the powder on the plate. It was as astonishing as if a beam of sunlight had suddenly broken through some very heavy curtains.

By 1900 all the naturally occurring elements with the heaviest atoms – polonium, radon, actinium, thorium, protoactinium and radium – had been found to be radioactive like uranium, and Ernest Rutherford and Frederick Soddy, working at McGill University in Canada, had discovered that their atoms were splitting up quite spontaneously with the emission of three different kinds of penetrating radiation which were called α-, β- and γ-rays.

It was a far cry from the old idea of atoms once described by a schoolboy as 'very small, hard marbles invented by Dr. Dalton' and regarded in the early days, even by Rutherford, as 'nice hard fellows'. A contemporary scientist put it in rhyme:

> So the atoms in turn, we now clearly discern,
> Fly to bits with the utmost facility:
> They wend on their way, and, in splitting, display
> An absolute lack of stability.[1]

The most immediate impact of the newly discovered phenomenon of radioactivity was the use of radium in medicine because the rays it emitted were very useful in treating cancer by what came to be known as radiotherapy. But Rutherford and Soddy also realized that radioactive materials provided a potential new source of power because they emitted considerable amounts of energy as their atoms disintegrated. It was not, however, easy to see how the energy could be used in practice because it was released so slowly.

Soddy[2] drew attention to the peaceful possibilities in a lecture on 'The Internal Energy of the Elements' in 1906, when he said 'that by the expenditure of about 1 tonne yearly of uranium, costing less than £1,000, more energy would be derived than is supplied by all the electric supply stations of London put together. One little step, so easily anticipated in imagination . . . divides us from this great inheritance.' 'We are starting the twentieth century', he wrote, 'with this prize in full view.'

Rutherford had expressed a much more alarming view when he wrote in 1903 that 'could a proper detonator be found, it was just conceivable that a wave of atomic disintegration might be started through matter, which would indeed make this old world vanish in smoke'.[3] He even went so far as to say, in jest, that 'some fool in a laboratory might blow up the universe unawares'.

'I'm not sure, Sir, but I *believe* I've split the atom.'

A Punch *cartoon by David Langdon, 1937 (Punch Library Services)*

These ideas were dramatized by H.G. Wells in his book, *The World Set Free*, published just before the start of the First World War. He vividly portrayed an international conflict, which he set in 1956, with the major cities of the world being destroyed by atom bombs. This was science fiction on a grand scale in 1914, yet it became something of a reality in 1944, some twelve years before Wells had prophesied.

* * *

The key to releasing atomic energy was the discovery of the neutron by James Chadwick in 1932. While working in the Cavendish Laboratory at Cambridge, under the direction of Rutherford, he had observed some very odd results when he bombarded beryllium with α-rays from radium, and he announced in a letter to *Nature* dated 17 February 1932, that he could only explain the results on the assumption that hitherto unknown particles with a mass similar to that of the hydrogen atom, were being formed. He called them neutrons because they were not electrically charged.

Their discovery, which had in fact been predicted by Rutherford as early as

Enrico Fermi. (Argonne National Laboratory)

1920, during his vastly important work in elucidating the structure of atoms, provided a source of particularly penetrating particles with which to probe the interior of atoms, and an Italian scientist, Enrico Fermi, undertook this task. He was born in Rome on 29 September 1901, the son of a high-ranking civil servant. He was a very clever boy with a keen interest in physics and by the time he had graduated from the *liceo*, a year earlier than was normal, he had read all the best-known books on physics and had said that he wanted to dedicate himself entirely to that subject. He continued his education at the University of Pisa and did original work in the field of relativity while he was still a student there before going on to work in Göttingen and Leiden. At the age of twenty-five he was chosen, through a national competition, to occupy the newly created post of Professor of Theoretical Physics at the University of Rome and, as there was no very strong tradition of research in physics in Italy, he was free to follow his own inclinations. He chose to investigate the effect of bombarding different elements with neutrons and he gathered a strong team of colleagues around him – Amaldi, D'Agostino, Rasetti, Segrè and Pontecorvo – whom he referred to as 'his boys'. They called him 'the Pope'. And what a field-day they had, publishing papers at the rate of about one a week until the end of 1938 in which year Fermi was awarded the Nobel Prize for his work.

The team found that the results of bombarding atoms with neutrons depended on the speed of the neutrons and on the size of the atoms. Sometimes the neutrons simply ricocheted off one atom on to another, rather like billiard balls struck into a conglomeration of heavy cannon balls. Sometimes the neutrons chipped small particles off the atoms they were bombarding, converting them into slightly smaller atoms of a different element. Sometimes the bombarded atoms captured neutrons and were converted into larger atoms. Because there were so many possibilities it was not at all easy to interpret all the experimental results and those obtained by bombarding uranium with neutrons were particularly perplexing. Fermi was convinced that at least some of the uranium atoms were capturing neutrons and being converted into larger atoms and, as the uranium atom was the largest known atom, this meant that atoms of some new element must be being formed. He reported the possible creation of such new elements in June 1934 and his conclusion, announced publicly in the presence of the King of Italy and well advertised in the world's press, was widely, if not universally, accepted.

Yet many discrepancies remained unexplained and the matter was not finally resolved until the end of 1938 when two Germans, Otto Hahn and Fritz Strassman, and two Austrians, Miss Lise Meitner and her nephew Otto Frisch, who had been doing similar work to that of Fermi, suggested that the neutrons were splitting at least some of the uranium atoms in two to make new atoms of approximately half the size. It was as if a marksman, accustomed to his bullets making small holes in his target suddenly found that they tore it in two. Hahn described his experimental observations as 'contradicting all previous results of nuclear physics',[4] and Frisch, who borrowed the word 'fission' from the biologists to describe the splitting of the uranium atoms, was so excited that he wrote to his mother saying, 'I feel as if I had caught an elephant by its tail, without meaning to, while walking through a jungle. And now I don't know what to do with it.'[5] More importantly, Frisch realized that such a fission of a uranium atom would release energy (it was rather like cutting a compressed spring in two or setting off a mouse trap) and he was able to calculate, and measure, the amount of energy that would be involved.

The news of the fission of uranium and the release of energy involved spread round the world like wildfire at the start of 1939, and all the excitement it generated was heightened by another discovery in the early spring. On 10 February Hahn and Strassman suggested that one of the 'side-effects' of the fission of uranium by a neutron might be the liberation of more neutrons, and this was confirmed by other scientists shortly after. The extra neutrons that were formed were called secondary neutrons.

There had been many breakthroughs in recent months but this was the big one because it was immediately apparent that if the fission of one uranium atom produced, say, three secondary neutrons, then each neutron might be able to split three more atoms, thus producing nine more secondary neutrons, and so on. A chain reaction, which could build up like a chain letter, might be possible so that just one neutron might lead to the fission of millions of uranium atoms. The release of energy would build up in geometric progression – 1 to 3 to 9 to 27 to 81 to 243 to 729 to 2,187 to 6,561 to 19,683 and so on.

*Leo Szilard. (Argonne
National Laboratory)*

Practical use of atomic energy was now in sight. If the energy could be released slowly, in a controlled way, a completely new source of power for peaceful purposes would be attainable. If it was released rapidly it might be possible to make a bomb.

That such things might be possible came as a great surprise to many scientists. Within recent years, Rutherford, Einstein and Bohr had all been pouring cold water on the idea. But it was no surprise to Leo Szilard, another of the many central Europeans who contributed so much to all the exciting developments. After an early education in Hungary, where he was born in 1898, he continued his training at the University of Berlin and he stayed in Germany until he felt forced to leave, in 1933, as a result of the anti-Semitism of the Hitler regime. After working in England for five years he emigrated to the United States in 1938. He applied his very considerable intelligence over a wide field, doing original work in thermodynamics as well as taking out a number of patents in such practical fields as domestic refrigeration and X-ray treatments. But in 1932, strongly influenced by the writings of H.G. Wells and interested in the implications of the discovery of the neutron, he became obsessed with nuclear physics and devoted the rest of his life to it. As early as 1934 he began to file a number of patents in London relating to 'the liberation of nuclear energy for power production and other purposes'. He had in fact foreseen the possibility of a chain reaction involving neutrons some time before it became a reality.

Szilard was also a great idealist with strong political views and somewhat utopian plans for changing, and even saving, the world. He fully realized the implications of the ideas in his patents so he assigned them to the British Admiralty so that they could be kept secret, after the War Department had initially spurned them. He was dismayed when he learnt in 1939 that his idea had become a real possibility, because he thought that it meant 'the world was headed for grief'.[6] To try to limit the damage he foresaw, he made strong efforts to ensure that the detailed information about the uranium chain reaction was kept secret. He was supported in this by a fellow Hungarian, Edward Teller, but the information was published by others, and events moved strongly against him.

On 16 March 1939 Hitler annexed Czechoslovakia and the imminence of war led scientists in many countries to direct their attention towards building an atom bomb and to warn their governments of the consequences. And these efforts were intensified after the start of the Second World War on 3 September.

* * *

In Britain, G.P. Thomson, who was Professor of Physics at Imperial College, London, applied to the Admiralty for a ton of uranium oxide on which to experiment, and warned the British government of the dangerous prospects ahead and of the desirability of trying to corner the only significant stocks of uranium ores which were in the hands of a Belgian company, Union Minière, operating in the Belgian Congo (now the Democratic Republic of Congo). Similar warnings were received by the authorities in Germany, Russia and Japan. In the United States, Szilard borrowed 225 kg of uranium oxide for experiments he had planned in collaboration with Fermi, who had emigrated from Italy in 1938 because his wife, Laura, was Jewish.

Szilard and Fermi were both fearful of the outcome of their activities and they asked Professor Albert Einstein, the best-known and most respected scientist of the day, to draft a letter to Franklin Roosevelt, the President of the United States, explaining the possible dangers. The letter was dated 2 August 1939, but Alexander Sachs, a friend of the President to whom the letter was entrusted for delivery, felt that Roosevelt was too busy to be bothered with it until he finally took it to him on 11 October, along with a summary of the situation which he had prepared himself. President Roosevelt responded by setting up an Advisory Committee headed by Dr Lyman J. Briggs, the Director of the Bureau of Standards, and including both Szilard and Fermi, together with military and naval representatives, as members. The Committee reported back to the President on 1 November, recommending 'adequate support for a thorough investigation' but there was no great sense of urgency and some of the non-scientific members were contemptuous of the whole idea. The Army man said that there was already a big prize on offer to the discoverer of a death ray and added that 'it is not weapons which win wars, but the morale of the troops'. Others, fortunately, had more foresight.

* * *

Establishing a chain reaction in uranium was still only a theoretical possibility, and any hope that it would be easy to do in practice was soon dashed when it was discovered that the two main isotopes in uranium, ^{235}U and ^{238}U, reacted differently to neutrons of different speeds because of the difference in their masses. Slow neutrons are very effective in fissioning ^{235}U atoms but they are simply captured by ^{238}U atoms without causing any fission. Faster neutrons will also fission ^{235}U atoms, though less readily than slower ones, but they are, preferentially, captured by ^{238}U atoms, again without causing fission. It does in fact require very fast neutrons to fission ^{238}U atoms.

If, then, a block of natural uranium is bombarded by a beam of neutrons, fission is caused in the ^{235}U atoms, predominantly by the slower neutrons, but the faster ones are simply captured by the ^{238}U atoms. It is, then, well-nigh impossible to establish a chain reaction in a block of natural uranium because the secondary neutrons are fast ones, so that they are unable to promote the chain, being captured by the predominant ^{238}U atoms before they have much chance of fissioning any other of the thinly spread ^{235}U atoms.

There were at first two possible ways of tackling the problem. The ^{235}U, which is easily fissioned, could perhaps be separated from the ^{238}U, or an attempt could be made to slow down the secondary neutrons. Somewhat later, another alternative arose because it was discovered that capture of neutrons by ^{238}U atoms formed a heavier atom of a hitherto unknown element of the type reported by Fermi in 1934. It was called plutonium and, like ^{235}U, it was readily fissioned by neutrons of any speed.

*   *   *

Frisch had been educated at the University of Vienna before going to work in Germany between 1927 and 1933. Then, like Szilard and many others, he was forced to leave and, after a short time in England, he eventually settled down in Copenhagen to work with Niels Bohr. In 1939, worried by the possibility of a German invasion of Denmark, he came once again to England to work with an Australian, Mark Oliphant, the Professor of Physics at Birmingham University. There he joined forces with another refugee, Rudolf Peierls, a wealthy German who had been born and educated in Berlin and had gone to work in Cambridge in 1933. Anticipating the Nazi purge, he chose to remain in England and became a naturalized British citizen in February 1940.

Frisch had succeeded in separating the ^{235}U from the ^{238}U atoms in natural uranium to some extent, and had come to realize that, though it was not easy, it might be possible to do it more fully. What then might happen if a piece of reasonably pure ^{235}U was bombarded by neutrons? One possible answer to the question was distinctly alarming because, as there are always neutrons floating around in the atmosphere which could bombard the ^{235}U, it might, perhaps, blow up spontaneously. But Peierls was able to calculate that it would only have a chance of doing that if the piece was above a certain critical mass or size.

He argued that so many of the secondary neutrons produced in a small piece of ^{235}U would escape from its surface before fissioning further atoms that a chain reaction could not be set up. But in a larger piece, above the critical size, there would be more chance of the secondary neutrons causing fission before they escaped from the surface so that a chain reaction could be established. It is rather like lighting a bonfire. If the pile of material is too small, too much heat escapes from the surface so that the whole bonfire will not ignite. It has to be above a certain size before it will all catch fire. Peierls was able to do a rough calculation to find out that the critical size for ^{235}U would be something like that of a golf ball.

These highly significant results were summarized in two reports which were submitted to Professor Oliphant. They were remarkably prescient, pointing out that a bomb made from 5 kg of ^{235}U would cause an explosion similar to that of several thousand tonnes of dynamite, and suggesting that the bomb could be constructed in such a way that two pieces of ^{235}U, each below the critical size and therefore safe, could be pushed together to make a piece above the critical size in order to set the bomb off. They also drew attention to the fact that there would be a good deal of what came to be called fall-out and that effective protection from such a bomb would be well-nigh impossible. 'The bomb', they wrote, 'could probably not be used without killing large numbers of civilians, and this may make it unsuitable as a weapon for use by this country.'[7] On the other hand they recognized that the only reply to any possible German atom bomb would be the deterrent effect of a similar weapon. This was early in 1940, and the thought of what the Germans might be doing could not simply be overlooked.

Oliphant was convinced that the 'whole thing must be taken rather seriously' and this was done by a small committee which was set up in April 1940 under the chairmanship of G.P. Thomson. To try to hide its activities it came to be called the MAUD Committee, the name originating in a rather quaint way. When Denmark was overrun by German forces, Niels Bohr sent a cable to Frisch which ended: 'Tell . . . Maud Ray Kent.' Everyone was at first mystified by this phrase and some suggested, rather fancifully, that it was a secret message, in the form of an anagram, that the Germans had 'taken radyum' from Copenhagen. Eventually it was realized that it referred to Maud Ray who had been governess to the Bohr children and who lived in Kent. But it had been so puzzling that MAUD was suggested as a good name to cover up any activities, though some soon took it to mean Military Applications of Uranium Disintegration.

By 15 July 1941 the Committee, though initially very sceptical, had reached the conclusion that building an atom bomb was practicable and that it would 'lead to decisive results in the war'. It predicted that a bomb containing about 12 kg of ^{235}U would have the devastating effect of about 2,000 tonnes of TNT and that the cost of building a plant to separate the ^{235}U and ^{238}U in sufficient quantities to make three bombs would be about £5,000,000 and could be in operation by the end of 1943. In spite of the enormous cost, the Committee recommended that work should continue with the highest priority.

Winston Churchill, the British Prime Minister, accepted the main recommendations of the MAUD Committee on 30 August, writing, 'although personally I am quite content with the existing explosives, I feel we must not stand in the path of improvement, and I therefore think that action should be taken . . .'.[8] He was encouraged in his decision by Lord Cherwell (formerly

Professor Lindemann), his chief scientific adviser, who was strongly in favour of moving ahead though he set the odds of success at only two to one against or even money.

A new division of the Department of Scientific and Industrial Research, known as the Directorate of Tube Alloys to hide its real activities, was set up. It was directed by Sir Wallace Akers, who was released by Imperial Chemical Industries, and it rapidly took all those who were already working in the field under its wing. The British die was cast.

*   *   *

Meanwhile, American scientists had also been investigating the exploitation of nuclear fission. Alfred Nier had separated ^{235}U from ^{238}U in February 1940, but the main American effort was directed not towards separating the isotopes so that a chain reaction could be set up in pure ^{235}U but towards trying to establish a chain in naturally occurring uranium by slowing down the secondary neutrons which were produced. The idea, thought up by Szilard and Fermi, was to use blocks of graphite in which a small lump of naturally occurring uranium was embedded. The blocks could then be built up in a lattice structure, which came to be called a pile. Most of the fast secondary neutrons emitted by the fission of any ^{235}U atoms in the pile would have to pass through graphite before they could impinge upon any other ^{235}U atoms, and the effect of graphite on neutrons is to slow them down; it is known as a moderator. In a small pile sufficient neutrons might escape from the surface before a chain reaction could be set up, but as the pile got bigger and bigger it would eventually reach a critical size, just as pure ^{235}U does. But because ^{235}U atoms are very thinly spread in natural uranium and because many of the neutrons would still be captured by ^{238}U atoms the critical size of a pile would be very large.

Once preliminary experiments, designed to establish the amounts and types of material required and how best to arrange them, had shown that the idea was feasible, Fermi set about building a pile in a double squash court under the west stand of Stagg Field Stadium in the University of Chicago. It was referred to as CP-1 (Chicago Pile number 1). Building began in May 1942. A bottom layer of blocks of pure graphite was covered with two layers of blocks containing lumps of uranium. Then came another layer of pure graphite followed by two layers of uranium-containing graphite – and so on. The structure was supported on a wooden framework, and arrangements were made to incorporate movable sheets of cadmium to act as control rods. Cadmium is a particularly good absorber of neutrons so that there could be no chance of any chain reaction developing when the cadmium sheets were within the pile. Withdrawal of the control rods from the pile would enable the neutrons to 'go into action'.

The pile would not, therefore, operate until it was larger than the critical size and until the control rods were withdrawn sufficiently. The pile was built up, layer by layer, in the shape of an enormous balloon lying on its side. It reached the critical size at layer 57 when it was 7.6 m long and 6.1 m across. It contained about 360 tonnes of graphite, uranium and uranium oxide but, with the control rods fully inserted, it lay dormant.

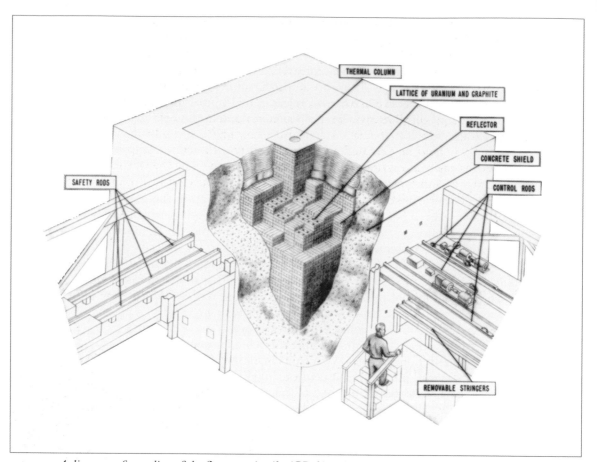

A diagram of a replica of the first atomic pile (CP-1) as it was rebuilt at Argonne. (Argonne National Laboratory)

It was the second day of a cold, snowy December when it was brought to life. After some preliminary checks in the morning, the fateful test began at 2 o'clock. Fermi directed operations from the balcony, originally intended for spectators of a squash match but now housing the panel of control instruments. An assistant stood by with a bucketful of concentrated cadmium nitrate solution to throw on top of the pile and absorb neutrons if anything should go amiss. Bit by bit, the last control rod was withdrawn until Fermi, with his eye firmly on the control panel, announced that the pile had gone critical. He allowed it to run for about 4½ minutes but only at a power output that would light a small torch bulb. The controlled release of nuclear energy had, for the first time, been achieved, and right in the heart of Chicago.

It had all been very unspectacular. Nothing had seemed to move and an eye-witness referred to an 'awesome silence'.[9] Another said that he had had an 'eerie feeling'.[10] But the general mood was best summed up by the one who said that 'he had seen a miracle'.[11] The phlegmatic Fermi raised a smile and those privileged to be present drank a little Chianti out of paper cups. But there was no toast and only muted elation. Szilard was, indeed, distinctly gloomy, saying that the occasion would 'go down as a black day in the history of mankind'.[12]

Yet it was a quite remarkable achievement. CP-1 had not been designed to produce great power. There were neither cooling arrangements nor any great shielding from radioactivity. But it had proved beyond any doubt that a nuclear chain reaction could be operated and Fermi said that the power output could be controlled 'as easily as driving a car'. It had, moreover, opened the way for making plutonium, for some of the ^{238}U atoms in the pile had captured many of the neutrons and been transformed into that new element.

* * *

The MAUD report, which had persuaded Churchill to support further action, recommended that the existing collaboration between Great Britain and the United States should be continued and strengthened. That was clearly desirable

Damage caused by the Japanese attack on the United States Pacific Fleet lying at anchor in Pearl Harbor, Hawaii, 7 December 1941. Left to right: *USS* West Virginia *(severely damaged);* USS Tennessee *(damaged);* USS Arizona *(sunk). (Imperial War Museum, NYF22545)*

but there was a certain lack of enthusiasm in the States and the Advisory Committee chaired by Lyman Briggs was not very influential. There were even suggestions, during the summer of 1941, that all work on nuclear fission should be dropped and Szilard had already become very impatient, saying that the project would never be completed if each step took ten months of deliberation. What was needed was less discussion and more decision. In the event it was only some vigorous lobbying by scientists working in Britain, particularly by the Australian Oliphant, who flew to the States in late August and by-passed Briggs – 'this inarticulate and unimpressive man' was how he described him – that tipped the scales. And they tipped quickly.

The MAUD report was officially delivered to the United States government on 3 October 1941, and within six days Roosevelt had appointed a Top Policy Committee to control events. He also wrote to Churchill on 11 October suggesting a closer collaboration between the two countries. That became more of a reality when the National Academy of Sciences submitted a report to the President, on 17 November, broadly supporting the findings of the MAUD Committee. If there had been any lingering doubts about the best course to follow they were swept away by the Japanese attack on Pearl Harbor on 7 December 1941. Within three days the United States was at war and all the resources on both sides of the Atlantic were committed to building an atom bomb. By then, too, it was obvious that the major part of the development would have to be carried out in the United States because British facilities and manpower were already stretched, and the country was under constant air bombardment by the German *Luftwaffe*.

CHAPTER 15

The Manhattan Project

The American research effort had been spread between a number of widely spaced universities and directed by individual professors. To draw these diverse activities and people together required a number of organizational and leadership changes, which eventually led to the appointment of Leslie Robert Groves as the director of the project in September 1942. The activities were camouflaged under the code-name of the Manhattan Engineer District, generally shortened to the Manhattan Project. Groves was a 46-year-old colonel in the Corps of Engineers who, as deputy chief of construction for the United States Army, was no stranger to supervising large-scale projects, such as the building of the Pentagon, and being responsible for spending vast sums of money. But he was anxious to get closer to the war zone and to command troops in action so he was not very pleased with his new appointment. He knew, however, where his duty lay and promotion to brigadier softened the blow and gave him greater authority for the formidable task that confronted him.

Groves was the son of a lawyer who had entered the ministry later in life and had served as a padre on the Western Front during the First World War. He stood about 1.8 m tall, but was distinctly overweight – estimates varied between 115 and 130 kg – and while all this flesh didn't enable him to present a very soldierly figure, particularly when wearing a belt, it did nothing to quench his dynamic energy or his ego. Nicknamed 'Greasy Groves' in his early days in the Army, he was not renowned for his tact, and was almost universally disliked; moreover, he was an Anglophobe. He was, nevertheless, widely respected and, by many, feared. His second-in-command said that he 'hated his guts'[1] and described him as 'the biggest sonofabitch he had ever met' but he still had confidence in his decisive, ruthless ability.

Groves had been trained in engineering at the Massachusetts Institute of Technology before going to West Point, where he was fourth in his class, but he was not enamoured of scientists and regarded most of them as inflexible, indecisive and argumentative. He did, however, sorely need someone to direct the new central laboratory where the bomb would be designed, and his choice fell on Robert Oppenheimer. It was a surprising, but inspired, selection.

Oppenheimer was tall and skinny, less than half the weight of Groves, and always gave the impression that he would benefit from a good meal and some physical exercise. His countenance was sad and lugubrious, his posture droopy, his movements puppet-like as though his joints were not well coordinated, and there was a general air of tension and nervousness about him which was exacerbated by his chain-smoking. He was born on 22 April 1904, the son of a German Jew who had emigrated to the United States in 1898 and who had prospered as an importer of textiles. Robert was a delicate child and was a distinctly late developer; he himself referred to 'an almost infinitely long adolescence'. He was tormented in his early years by illness, by a lack of self-confidence and by loneliness. He was never quite able to come to terms with himself or with the world, and he had an ascetic, poetic vision which was so unrealistic that he was always disappointed by any of his achievements. He was a perfectionist who was so clever and ambitious that his goals always seemed beyond reach. He expected everything and everyone to be extra-special, but they never were.

In his early days, then, first at Harvard and later at Cambridge and other European universities, he was distinctly unhappy. A psychiatrist indeed diagnosed (wrongly as it turned out) a type of schizophrenia, and Oppenheimer himself wrote that he was 'on the point of bumping myself off'. But he worked his way through all these difficulties and blossomed forth in his thirties, by which time he had turned his attention to studying the quantum theory and had begun to establish an international reputation in that field. He held joint professorial posts at the California Institute of Technology and the University of California where he established a world-famous school. He could still use a caustic tongue, still didn't suffer fools gladly and could still appear to be very arrogant. His private wealth also enabled him to act very independently, but he had, by and large, learnt to contain what he himself called his 'beastliness', so much so that his strong personality and his charm began to break through more and more clearly. They had a particularly marked effect on his students and on the opposite sex, and he began to gain a reputation as an inspired teacher and something of a lady-killer. Leona Bibby (later Marshall), who was one of the very few women actively engaged in the research work, described him as 'a disembodied spirit who never expressed himself completely and left a feeling that he had depths of sensibility and insight in layer after unrevealed layer like the seven veils of Salome'.[2]

Without doubt, Oppenheimer was a complex character with many of the drawbacks commonly associated with a touch of genius, and not everyone thought that Groves was wise to appoint him to direct anything. He had never led a very large group of people, his interests lay in somewhat abstruse theories, he was not a good experimentalist, he was regarded by many as indecisive, and he was not a Nobel Prize winner unlike many of the people with whom he would have to work. More importantly, his political leanings were to the left and he had been associated with members of the Communist party. His first fiancée, Jean Tatlock, whom he began to court in 1936 and with whom he had a stormy, on-and-off relationship, was a member of the party, and Katherine Puening (Kitty) whom he married as her fourth husband, in 1939, had been previously married to a member who had been killed fighting in Spain. Robert's brother, Frank, a fellow physicist, and his wife, were also former members of the Communist party. It was, however, Oppenheimer's friendship, since 1938, with Haakon Chevalier, a Professor of Romance Languages at the University of California, that was to prove the most damaging because, towards the end of 1943, Oppenheimer named Chevalier as the man who had been involved, albeit unsuccessfully, in an attempt to introduce him to a Russian agent. To try to protect Chevalier, Oppenheimer had originally prevaricated when tackled on this issue by the security authorities and he only revealed the full truth when pressurized by Groves to do so.

Groves admitted that such a background was 'not to our liking' but he regarded Oppenheimer so highly that he used his considerable authority to obtain his security clearance, writing, on 20 July 1943: 'In accordance with my verbal directions of 15 July it is desired that clearance be issued for the employment of Julius Robert Oppenheimer without delay, irrespective of the information which you have concerning Mr. Oppenheimer. He is absolutely essential to the project.'[3]

Though Oppenheimer said that he had never regarded Communist ideas as making any sense, the tag of being a fellow-traveller stuck with him and he was kept under constant surveillance and frequently interrogated. In the end his past, rightly or wrongly, caught up with him and his loyalty or lack of it became a *cause célèbre* in the postwar years. Nor did the two main women in his life, who were partially responsible for the tag, bring him constant happiness. Jean Tatlock committed suicide in 1944, and though Kitty bore him a daughter in that same year, she reacted to the general strain by turning to drink. But whatever the crosses he had to bear, Oppenheimer was to repay fully the faith that General Groves had showed in him, and the two of them, so unlike in many ways and only formally friendly, struck up a surprisingly good working relationship as they set about their task.

Roosevelt had decreed that 'time was of the essence', there was no shortage of funds and General Groves was no slouch. It was in his nature to act first and ask the questions later. He began by touring the universities where work was already going on, and by taking steps to ensure that adequate supplies of uranium would be available, while Oppenheimer began to recruit a team of scientists. The general plan was that some basic research would continue in some of the universities but that new centres would be built where the necessary personnel and facilities could be concentrated. In September 1942 a site was bought at Clinton in Eastern Tennessee, where ^{235}U would be separated from natural uranium; this came to be known as Oak Ridge. Another site, for making plutonium, was purchased in January 1943 at Hanford on the banks of the Columbia River in the state of Washington. The third, acquired in November 1942, at Los Alamos in northern New Mexico, was to provide a central laboratory, which would be directed by Oppenheimer himself, where the bomb would be designed and fabricated once the necessary materials became available. It was like building three new townships. The sites chosen were remote and far apart from each other to limit any attempts at sabotage and to provide good security. Groves hoped to compartmentalize all the work so that one team would not know much about what another was doing, with only very few people having an overall picture. This was a typically military approach and might have been good for security but it caused a lot of friction between the military and the scientists who were accustomed to much more open discussion. Secrecy was indeed to be a contentious issue throughout.

The gaseous diffusion plant at Oak Ridge where ²³⁵U was extracted from natural uranium. The plant was four storeys high and covered an area of 250 sq. km. (Imperial War Museum, RNY75473)

To push ahead as rapidly as possible the method of parallel development was adopted. If more than one way of solving a problem seemed to be feasible they were all investigated fully without a lot of preliminary consideration of the detailed pros and cons. This was expensive and inevitably led up some blind alleys but enough clear pathways opened up to show the way forward.

* * *

The Oak Ridge site, where ²³⁵U was to be extracted from natural uranium, covered 250 sq. km of impoverished, sparsely populated, rural land consisting mainly of red clay which meant that there was generally a lot of mud about when it was wet. Initially, a new town to accommodate 13,000 people was planned but a peak population of 75,000 was reached, during the summer of 1945, making Oak Ridge the fifth largest city in Tennessee. It was a closed military area, surrounded by barbed wire, and all the services associated with a large town – housing, water, drains, electricity, schools, hospitals, churches, roads and railways – had to be provided, as well as all the special technical buildings and plant.

Although ^{235}U had been separated from ^{238}U on a small laboratory scale, the task of doing it on a large industrial scale was Herculean. Natural uranium contained only 0.7 per cent of ^{235}U, and it differed in mass from ^{238}U by only 1.26 per cent. One worker said that it 'was like trying to find needles in a haystack whilst wearing boxing gloves'.[4] Another likened it to a blind man trying to find one black sheep among 140 white ones.

There was too little know-how for anyone to predict which would be the best method of tackling the problem so at first two methods were adopted: electromagnetic separation and gaseous diffusion; later, a third method, thermal diffusion, was also tried. All three processes took bold steps into unknown territory, and because of the lack of time, research, development, construction and operation, all had to go on more or less simultaneously. In such a rapid build-up of new technology it is not surprising that there were formidable hurdles to overcome. The recruitment and training of staff; the procurement of all the vast amounts of necessary materials; the design and building of the plant; and the successful operation of the new processes, all provided many headaches and many setbacks.

In the electromagnetic method of separation, which had been developed mainly by Ernest Lawrence at the University of California, the vapour of a uranium compound is electrically charged and then passed through a vacuum surrounded by very strong, 4,570 tonne, oval-shaped magnets. The magnetic field deflects the electrically charged uranium atoms, but it deflects the lighter ^{235}U ones more than the heavier ^{238}U ones, so that the two are separated. It is rather like the effect of a cross-wind on moving cars; the lighter cars are thrown further off course than the heavier ones. The set-up was known as a calutron or racetrack, and some 500 of them were planned to provide the right amount of ^{235}U. Construction began in February 1943, and the plant was first operated in November of that year, but, by the end of the year, there had been so many faults in the magnets that it had scarcely produced 1 gram of ^{235}U. It was not until September 1944, after all the original magnets had been stripped down and rebuilt, that more promising amounts began to be made.

In the gaseous diffusion method, natural uranium was converted into uranium hexafluoride (hex), which is a gas consisting of a mixture of ^{235}UF$_6$ and ^{238}UF$_6$ molecules. When the gas was pumped through a porous filter or barrier, the lighter ^{235}UF$_6$ molecules passed through slightly more rapidly than the heavier ^{238}UF$_6$ ones, so that the gas emerging from the filter was enriched in ^{235}UF$_6$. Only very slight enrichment was achieved by one filter so that the process had to be repeated over and over again through a cascade of thousands of filters. The main problems were the handling of the very corrosive uranium hexafluoride gas and the manufacture of suitable filters. All the hundreds of kilometres of pipes in the plant had to be plated with nickel to withstand the gas, and a special, new sealant (now widely used under the trade name Teflon) had to be developed for use in the thousands of pumps because hex attacked any normal greases. The difficulty in making satisfactory filters arose because they had to have very fine pores and yet be strong enough to withstand the gas pressures to which they were subjected. The first type of filter was replaced by an improved design, but this caused some

delay so that it was not until early in 1945 that the gaseous diffusion plant began
to produce reasonable amounts of enriched uranium.

In the thermal diffusion process, isotopes are separated by feeding the mixture
into a vertical, cylindrical tube, which is cooled on the outside and which has a
heated, concentric cylinder inside along its axis. The lighter isotopes move to the
hot, inner surface and then to the top of the cylinder. The heavier ones move to
the outer, cold surface and downwards to the bottom of the cylinder. One cylinder
produces only very little separation but racks of thousands of cylinders achieve
the desired results. Construction of the plant began in June 1944 and, despite
many problems with leakages, it was operating reasonably well at the start of
1945.

It might seem profligate to operate three different processes simultaneously,
but, in the end, they all earned their keep because it was only by using the
partially enriched uranium from one process as the feed-stock for another that the
required product was made in sufficient quantities. Planning was exceptionally
difficult because, particularly in the early stages, neither the actual amount of ^{235}U
required for a 'Little Boy' bomb, nor the desired degree of purity, were known
with any certainty. Estimates of the amount began at around 100 kilograms but, in
the end, around 40 kilograms of ^{235}U of about 80 per cent purity was found to be
satisfactory, and slightly more than this amount was delivered to Los Alamos by
24 July 1945. It was less than three years since General Groves had arranged for
the purchase of the Oak Ridge site.

*   *   *

To make plutonium, which was required for 'Fat Man', it was necessary to build a
much larger version of the pile that Fermi had constructed in his Chicago squash
court. Some development work was carried out in the Argonne Forest, close by
Chicago, and at Oak Ridge, but the final production site which was chosen was at
Hanford in the state of Washington on the banks of the Columbia River. The
availability of large supplies of water from the river for cooling purposes, and the
proximity of the Grand Coulee Dam with its associated hydroelectric scheme,
made the site very satisfactory. The original Hanford was a tiny, riverside village
with a sparsely populated hinterland of storm-ridden, scrubby desert. General
Groves inspected the site on 16 January 1943 and about 2,000 sq. km of land was
acquired in February. As at Oak Ridge, a new town was spawned and, at its peak,
it had a population of about 60,000.

The manufacture of plutonium involved two stages. First, it had to be made
within a pile, or a reactor as it was sometimes called, by the bombardment of ^{238}U
by neutrons. But as only a small part of the ^{238}U was converted into plutonium,
this had to be followed by a chemical extraction process to obtain pure plutonium.
Three piles were built, at 10 km intervals, alongside the Columbia River. These
were on an altogether different scale from Fermi's original one: he called them
'different animals'. Not only were they much bigger, they also used purer
uranium amd graphite and they incorporated extensive cooling and shielding
arrangements so that they could be run at much higher power. Thousands of

One of the separation plants for extracting plutonium at Hanford, built and operated by DuPont. The huge 'box' was known as a 'Queen Mary'. (The Hagley Museum and Library)

aluminium tubes, embedded in cylinders of solid graphite, 8.5 by 11 m in size, contained slugs of uranium, themselves clad with aluminium. When the pile was operating, the slugs were kept cool by passing water over them through the tubes. At the peak, 350,000 litres of water per minute were used, the heated water being returned, after treatment, into the river. To avoid ill-effects from the extensive radiation which was emitted, the pile was shielded by surrounding it with steel plates, concrete, and special high-density compressed wood.

After the pile had operated for about seven weeks, the slugs of uranium, now containing about 250 parts per million of plutonium, were simply pushed out of the aluminium tubes into tanks of water and replaced by new slugs. After being stored in the water for about eight weeks to allow time for the short-lived radioactivity to decline, the slugs were ready for chemical treatment to extract their small plutonium content. The extraction process, which had been worked out by Glenn Seaborg, involved some complex chemistry and, because the slugs were still highly radioactive, some complex technology. The process, beginning with the solution of the slugs in hot, concentrated nitric acid, was carried out in three vast plants erected 16 km away from the piles. Each plant consisted externally of massive concrete structures, 243 m long, 20 m wide and 24 m high. The walls were made of 2.1 m thick concrete and the roof was 1.8 m thick. They were called 'Queen Marys', after the huge ship of that name. There were various cells within each 'Queen Mary' in which the different chemical operations were carried out, but the radiation was so intense that they all had to be done by remote control through the concrete walls, using periscopes, television cameras and automatically operated handling equipment.

Construction of the first pile began in August 1943, and it was first operated in September 1944. The other piles were also in action by the end of 1944 and the first supplies of plutonium to Los Alamos were delivered at the start of 1945. By the middle of that year enough had been provided to make two 'Fat Man' bombs, and much more was in the pipeline.

∗   ∗   ∗

Los Alamos was set among rugged, forested country in an isolated region of extinct volcanoes. The whole area was split up by deep canyons into long, thin plateaus, called *mesas*, which were often at a height of 2.3 km. Access to the *mesas* was extremely difficult and the area could only be reached after a hazardous, 56 km drive from Santa Fe, the nearest community of any size. When the site was first inspected by Groves and Oppenheimer, it was occupied only by a few ranchers and by a private school for boys, and it was the school buildings that provided a base from which to expand. Oppenheimer was particularly happy with the choice of site because he knew the area well, having shared a cabin some 100 km away with his brother for a number of years.

Los Alamos was officially opened on 15 April 1943, but many of the planned, new buildings were far from ready and there was a general air of chaos and confusion as Oppenheimer set about getting the show on the road, travelling all over the country recruiting 'the men who could make a success' of this venture into the unknown. It was not easy to persuade scientists who were already fully occupied in jobs of their own choice to move to a place they had never heard of to work on a project, closely linked with the military, about which most of them knew very little. 'There was a great fear', Oppenheimer wrote, 'that this was a boondoggle which would in fact have nothing to do with the war we were fighting.'[5] But in the end his persuasive powers were such that he succeeded in gathering together what Groves described as 'the greatest collection of eggheads ever'.[6]

It was an international cast, with many native Americans supported by others whose origins lay in Italy, Hungary, Poland, Austria, Germany and Russia. As G.P. Thomson wrote: 'It is noteworthy, and I hope will be noted by future

An aerial view of Los Alamos. (Los Alamos National Laboratory)

Dictators, how dominating was the part played by physicists who had fled from Fascism and Nazism.'[7] And at the end of 1943, after an agreement on the mutual exchange of atomic energy secrets reached between Roosevelt and Churchill at a conference in Quebec, they were joined by many British scientists from the 'Tube Alloys' project. Akers, Frisch, Peierls, Chadwick, Oliphant, Penney, and Niels Bohr, with his son, Aage, were among those who went to work across the Atlantic. So, too, was Klaus Fuchs, a German who had become a naturalized British citizen, and who slipped through all the security screens to become infamous for passing secret information to the Russians.

Much basic work had to be done on the chemistry and metallurgy of uranium and plutonium and on how they could be handled safely, but the major problems that had to be resolved were the critical sizes of the two materials and the details of how they could be made into a workable bomb. The critical masses were measured in a series of dangerous experiments in which subcritical masses of the materials were brought together into a supercritical mass for a fraction of a second. The necessary work was carried out by what was known as the Critical Assembly Group operating in a building in a canyon remote from the main laboratories. In one set of experiments a piece of uranium hydride enriched with ^{235}U was dropped through a ring of the same material. As it passed through, the assembly became supercritical for a fraction of a second, simulating a small bomb, and it was possible to measure the output of energy. Otto Frisch, whose idea it was, referred to the experiments as 'twisting the dragon's tail'. Other experiments involved building small blocks of the materials into a small pile until the critical size was just reached, and it was such an operation that caused the only wartime death at Los Alamos. Harry Daghlian, working alone at night contrary to agreed regulations, accidentally dropped one of the blocks into the pile and the resulting surge of radiation led to his death twenty-four days later. A fellow worker, Louis Slotin, was killed by a similar incident after the war.

The first idea for making a bomb was to assemble two subcritical pieces of material at opposite ends inside a sealed gun barrel. To set off the bomb, one of the pieces would be fired into the other, using an explosive charge, to form a supercritical mass and, at the same time, to trigger off a source of neutrons. The two pieces must come together at just the right speed; if it happens too quickly they might just disintegrate on impact before the chain reaction could get started; if too slowly, the chain reaction might build up too slowly and simply produce enough heat to cause the mass to expand without exploding.

In the event, this 'gun' method was found to be satisfactory for the ^{235}U used in 'Little Boy' but it would not work for plutonium as two pieces of that material could not be fired together quickly enough in a gun barrel. This was due to the presence of too high a proportion of ^{240}Pu atoms among the predominant ^{239}Pu atoms. It was therefore necessary to use a much trickier implosion method in making 'Fat Man'. The idea was that two touching hemispheres of plutonium, forming a sphere which was below the critical size and which had a neutron source at its centre, would be surrounded by explosive charges which, when detonated, would compress the sphere into about half its original size. The corresponding increase in density of the plutonium would mean that the atoms were brought closer together so that the neutrons responsible for carrying on a chain reaction would have to travel much shorter distances before colliding with other plutonium atoms. The smaller sphere of high-density plutonium would, therefore, be supercritical.

The necessity of developing two rather different types of bomb – the 'gun' for ^{235}U and the 'implosion' for plutonium – meant that the research effort had to be two-pronged and this threw still greater burdens on Oppenheimer's organizing ability and severely stretched the facilities at Los Alamos. There were also very difficult decisions about priorities for, at that stage, it seemed likely that ^{235}U, which could be made into a bomb relatively easily, would not be available as early as plutonium, which was clearly going to be more difficult to make into a bomb. Nor was the situation helped when Seth Neddermeyer, a fine scientist but a poor organizer, who directed the original work on the implosion method, had to be replaced by George B. Kistiakowsky because progress was being made too slowly.

It was an almost completely new field of research which, because of the lack of any basic, background work, involved a good deal of imaginative, intuitive guesswork. The idea was to surround a central plutonium sphere with a layer of natural uranium and to have an outer layer of specially designed, shaped explosive charges. These charges would concentrate the explosion towards the centre of the sphere and as this concentrating effect was similar to that of a lens on light the charges were referred to as lenses. The lenses, which had to be made with a precision of a few thousandths of a centimetre so that they would fit together accurately enough, were cast in moulds and finished by machining. There were two components. One contained a fast-burning explosive mixture, called Composition B, consisting of TNT, RDX and a wax. The other was a slower-burning mixture, called Baratol, containing TNT, nitrocellulose, barium nitrate and aluminium powder. The two together were intended to build up a smooth, symmetrical detonation wave which would first compress the natural uranium layer and then the central plutonium. Within a period of about two years, over 20,000 castings and 50,000 machining operations were carried out to try to achieve the desired results, but, after all the tests, there remained some lingering doubts as to whether the radically new design for 'Fat Man' would work satisfactorily. It was therefore decided, during 1944, that any uncertainty should be resolved by carrying out a full-scale test firing on a prototype 'Fat Man'. The decision was made easier in the knowledge that there was expected to be enough plutonium available by the summer of 1945 to make two, or even three, bombs. It was lucky that similar testing of the simpler 'Little Boy' bomb was not regarded as essential, because the supply of ^{235}U was distinctly limited.

* * *

The test of 'Fat Man' was carried out in the south of New Mexico on a 30 by 39 km site on the edge of the Alamogordo bombing range. It was an area of flat, desert scrubland in the centre of a sun-drenched valley, so dry and hot during the day that it had been christened Dead Man's Trail by early Spanish explorers. The test and the site were code-named Trinity, a suggestion originating from Oppenheimer's imaginative mind and inspired by his recollection of lines from one of John Donne's 'Holy Sonnets':

> Batter my heart, three person'd God; for you
> As yet but knock, breathe, shine, and seek to mend.
> That I may rise and stand, o'erthrow me, and bend
> Your force to break, blow, burn and make me new.[8]

The new site, like those at Oak Ridge, Hanford and Los Alamos required the creation of still another new township from nothing.

The test bomb was to be exploded at the top of a 30 m high metal tower, built on a spot nicknamed 'Ground Zero'. Between 8 and 10 km away, to the north, west and south, there were observation posts built of earth with roofs of concrete slabs supported on strong wooden beams. The base camp was situated 6.5 km

Replicas of the two atomic bombs. 'Little Boy' (above) and 'Fat Man'. (Imperial War Museum, MH6809 and MH6810)

further to the south, and there was a viewing platform for VIPs on a hill 32 km from Ground Zero.

After some postponements because the bomb was not ready, 16 July 1945 was chosen as the fateful day, and Groves and Oppenheimer, together with other senior scientists and distinguished visitors, arrived on the 15th. By then the bomb had been largely assembled on the top of the tower, but not without much difficulty because a desert hailstorm, with high gusts of wind and much thunder and lightning, had severely hindered the work. The poor weather continued to cause much concern so that the planned timing of the test was changed first from 2 a.m. to 4 a.m. and then to 5.30 a.m.

It was a long, almost intolerable wait, which seemed like an eternity, particularly for the majority who had no active part to play in the final preparations. Some managed to snatch some sleep, but many just wandered around rather aimlessly in the dark. Groves tried to keep everyone calm, and was particularly helpful to Oppenheimer, but as the moment of truth approached the general tension mounted. As General Thomas F. Farrell, Groves's deputy, put it:

> We were reaching into the unknown and we did not know what might come of it. It can be safely said that most of those present were praying – and praying harder than they had ever prayed before. If the shot were successful, it was a justification of the several years of intensive effort of tens of thousands of people – statesmen, scientists, engineers, manufacturers, soldiers, and many others in every walk of life.[9]

Observers had been told to smear their skin with anti-sunburn cream, to close their eyes and cover them with their hands, and to lie down on the ground, or in a shallow trench, with their backs and feet to the tower. A green Very rocket was fired at 5.25 a.m.; then there was another 1 minute warning rocket; and, with only 45 seconds to go, an automatic timing system came into play which was under the control of Donald Hornig, who had designed the mechanism which was to fire the bomb and who had been the last man down from the top of the tower after making the final adjustments to the bomb. He, alone, could still have stopped the test at any moment by pressing a knife switch, but it was not necessary, and the first atom bomb in the world exploded 15 seconds before 5.30 a.m. on 16 July 1945.

It was still dark and the first effect was a burst of brilliant, searing light which lit up the whole area far more brightly than the midday sun. The flash was seen 290 km away, and the closer observers sensed the enormous intensity of the light even through their closed, covered eyes. They also felt the heat as though the door of a very hot oven had been opened. As they turned to see what had happened, using smoked glass to look through, they saw an enormous ball of red, orange and yellow flames of fire rising upwards out of the desert. Almost immediately they felt a blast of air, which knocked over some of those who were standing up, and this was followed by the thunderous roar of the explosion. Thereafter, the fireball developed, quite slowly, into a mushroom-shaped cloud of dust, which rose to a height of almost 11 km before being blown in many different directions by the variable winds at different altitudes.

Robert Oppenheimer (left) and Leslie R. Groves at the site of the Trinity test after the explosion. (Los Alamos National Laboratory)

The explosion made a crater 350 m in diameter in the desert floor, and, immediately below the tower, which had all but disappeared, there was another hole 36 m wide and 2 m deep filled with a glass-like substance because the temperature had been high enough to melt the desert sand. A 10 cm iron pipe, 450 m from Ground Zero had evaporated; 800 m away, a 21 m high tower built of 40 tonnes of steel set in concrete foundations, was flattened and ripped apart; and plate-glass windows 300 km away were cracked. Measurements showed that the bomb had been equivalent to almost 19,000 tonnes of TNT, which was almost four times more powerful than the Los Alamos designers had expected. What those present had seen and felt was quite unprecedented and they found it difficult to put their observations or emotions into words. Sensible comment and discussion immediately after the event, was not really possible, because everyone was stunned and giddy. Some found the sheer elation that their efforts had been crowned with success difficult to cope with; others were overcome by what Dr Kenneth Bainbridge, the director of the test, called 'this foul and awesome display',[10] yet others were already beginning to be fearful of what they had unleashed.

General Groves maintained a cool, military composure, telling General Farrell that the war would be over after two bombs had been dropped on Japan, and he congratulated Oppenheimer, saying, quietly, 'I am proud of all of you.' Oppenheimer replied with a simple 'Thank you',[11] but his deeper thoughts were perhaps better expressed in a line he recalled from the Hindu *Bhagavad-Gita*: 'Now I am become death, the shatterer of worlds.'[12] Bainbridge echoed the sentiment by saying that 'they were all now sons of bitches',[13] and even the ice-cold Fermi was so affected that he had to ask a friend to drive him home. General Farrell summed up much of the general mood when he wrote '. . . the strong, sustained, awesome roar which warned of doomsday and made us feel that we puny things were blasphemous to dare to tamper with the forces heretofore reserved to the Almighty'.[14] And he was probably even more in tune with reality when he added: 'Words are inadequate tools for the job of acquainting those not present with the physical, mental and psychological effects. It had to be witnessed to be realised.'

*   *   *

The success of the Trinity test in 1945 had a catalytic effect on events. In Europe Adolf Hitler had committed suicide, the German armies had surrendered in the early hours of the morning of Monday 7 May, and General Eisenhower, the Supreme Allied Commander, had announced in a radio message on 8 May – VE-Day – that 'the sounds of battle have faded from the European scene'. It was estimated that around forty million lives had been lost – directly or indirectly – in the six year carnage. Ironically, and surprisingly, it transpired that the idea of a German atom bomb, which had been such a driving force behind the British and American efforts, was but a myth. The Germans had never undertaken any serious, long-term project to make such a bomb even though they were well aware of the possibilities. In the first two years of the war, a group of scientists, led by

Werner Heisenberg and known as the Uranverein (the uranium society), had made considerable progress in devising methods of separating uranium isotopes and in making an experimental atomic pile, but in the middle of 1942 it had been decided to direct the main research effort into flying bombs and rockets.

Unhappily, President Roosevelt did not live to see VE-Day. He had collapsed, while sitting for a portrait, on 12 April, and died the same day. He had served as President for thirteen years and contributed so much to the successful outcome of the European conflict.

But the war in the Far East was far from over. The Japanese had overrun many far-flung islands in their early, successful days when they almost reached Australia and, when the counter-attack took place, they fought to the death to maintain their hold on them. Recapturing the tiny island of Iwo Jima cost the Americans over 4,000 dead, while at Okinawa over 12,500 Americans and 100,000 Japanese were killed in twelve weeks' fighting. Estimates suggested that the Americans might suffer over 200,000 casualties if they went remorselessly on their way and attempted a direct assault on the Japanese home islands.

The new President, Harry Truman, had to face up to this challenge. It is a measure of the level of security surrounding the Manhattan Project that, although he knew of its existence, he knew nothing of its purpose. So, at the same time as he was being briefed about the atom bomb, he was having to begin to decide what to do with it. All the arguments, some of which seem more powerful now than they did at the time, were expressed with great vehemence, and few national leaders can ever have been confronted with such an agonizing decision. In the event Truman decided, though not very positively, around 1 June that the atom bomb should be used against Japan. Groves remarked that 'he did not so much say "Yes" as not say "No". It would indeed have taken a lot of nerve to say "No" at that time.' After all, nearly two billion dollars had already been spent on the Manhattan Project so it could not be discarded lightly. British approval for the use of the bomb, which was a necessary part of the Quebec Agreement, was given on 4 July. Churchill's attitude was distinctly pragmatic. He wrote:

> There never was a moment's discussion as to whether the atomic bomb should be used or not. To avert a vast, indefinite butchery, to bring the war to an end, to give peace to the world, to lay healing hands upon its tortured peoples by a manifestation of overwhelming power at the cost of a few explosions, seemed, after all our toils and perils, a miracle of deliverance.[15]

Later in July Churchill and Truman met Stalin in conference at Potsdam, and Truman, in full knowledge of the success of the Trinity test, took the opportunity of casually informing the Russians of the American intention of using a special weapon to force a Japanese surrender. Stalin showed no great surprise or interest because, unknown to the Americans or British, Karl Fuchs had been feeding secret information to the Russians for many years, though his activities did not come to light until 1950 when he was back in England as head of theoretical physics at the Atomic Energy Research Establishment at Harwell.

At the end of the Potsdam Conference, on 23 July, an ultimatum on behalf of

The Potsdam conference, July 1945. Left to right: *the British Prime Minister, Winston Churchill; the American President, Harry Truman; the Soviet Premier, Joseph Stalin. (Imperial War Museum, KY74856)*

the governments of the United States, the United Kingdom and China was issued to Japan. It became known as the Potsdam Declaration and it laid down the conditions under which the war could be ended, concluding as follows: 'We call upon the government of Japan to proclaim now the unconditional surrender of all Japanese armed forces, and to provide proper and adequate assurances of their good faith in such action. The alternative for Japan is prompt and utter destruction.'[16] General Groves had already drafted a directive to General Carl Spaatz, the Air Force Commander, on 23 July. It began: 'The 509 Composite Group, 20th Air Force, will deliver its first special bomb as soon as weather will permit visual bombing after about 3 August 1945, on one of the targets: Hiroshima, Kokura, Niigata and Nagasaki.'

The Japanese government decided to fight on; Colonel Tibbetts's men did their duty; and the nature of warfare – and, even, of life – was never to be the same again.

CHAPTER 16

Nuclear Fusion

It is one of the undeniable facts of history that weapons of war become remorselessly more effective with the passage of time. That is what happened with conventional explosives, and so it was, also, with atom bombs. It did not take long after the end of the war for the team at Los Alamos to improve on the earlier designs of 'Little Boy' and 'Fat Man', with particular emphasis being placed on the latter because it was regarded as the more efficient. By 1948 bombs more than twice as powerful had been tested, and the 2-times had increased to 25-times by 1952. But even these monsters were dwarfed by a new breed of atom bomb depending not on the fission of large atoms such as uranium and plutonium but on the fusion, at very high temperatures, of small atoms such as the isotopes of hydrogen. They were called fusion, thermonuclear, super or hydrogen bombs. That they could be made, and would work, was shown by the successful test of 'Mike', designed and built at Los Alamos, on 1 November 1952. It was far too heavy and cumbersome to be regarded as a 'deliverable' bomb but it was the forerunner of what was to come.

'Mike' was shipped to the atoll of Eniwetok, one of the Marshall Islands, which were supervised by the United States under a United Nations' trusteeship, aboard the USS *Curtis*, and then taken by barge for final assembly on the small island of Elugelab. During the night before the test, as final preparations for the explosion were made, all the adjacent islands were evacuated, a final roll-call was taken, and the ships carrying most of the observers withdrew to a distance of 64 km. The sight they saw was distinctly unnerving, even to those who had witnessed the Trinity test. There was a ball of fire 4.8 km in diameter and the millions of litres of sea-water which had been turned into steam appeared as a great bubble. When visibility was restored it was clear that the island of Elugelab had been removed from the surface of the map. In its place there was a vast crater, 0.8 km deep and 3.2 km wide, in the reef. 'Mike' had exploded with a power equivalent to over ten million tons of TNT. It was 500 times more powerful than the 'Fat Man' dropped on Hiroshima and its blast would have demolished all five boroughs of New York. What Oppenheimer once called 'the plague of Thebes'[1] had left its first indelible mark.

* * *

Edward Teller is commonly referred to as 'the father of the hydrogen bomb', though he does not like the description and regards the whole enterprise as 'the

The 'Mike' test at Eniwetok in the Marshall Islands on 1 November 1952. **Above:** *before the blast. The bomb is in position on Elugelab, the fourth island from the bottom.* **Opposite:** *after the blast. Elugelab has disappeared, and the other islands are devastated. (Los Alamos National Laboratory)*

work of many people'. He was nevertheless the bomb's main advocate and it was largely his single-minded tenacity of purpose that eventually led to the bomb being built. He was another in the long line of Jewish refugees from Hungary who contributed so much to physics. Teller was born in Budapest on 15 January 1908, the son of a prosperous lawyer. He began his education at the Institute of Technology in Budapest, but, after living through his early teens in the turmoil following the 1918 revolution in Hungary, he left for Germany to study at the universities of Karlsruhe, Munich, Leipzig and Göttingen. When Hitler came to power in 1933, Teller felt forced to move once again, going first to Copenhagen and London, and then, in 1935, to the United States, where he was appointed Professor of Physics at the George Washington University. In 1941, shortly after

becoming an American citizen, he moved to Columbia University to work alongside Fermi and Szilard.

Teller, now an octagenarian, is of stocky build, with unruly hair and shaggy eyebrows, a long nose and bright blue eyes. He walks with a slight limp caused by an accident to one of his feet and his clothes are almost always rumpled and ill-fitting. He enjoys music and poetry; has always had a basically warm and friendly nature and a very happy family life; but, in his early days, he was ambitious, temperamental and argumentative so that he did not suffer fools gladly and did not always keep his friends. Even as a child, he was passionately interested in science and, later, he was particularly attracted by the broad aspects of new, way-out ideas and didn't care for organization or tedious detail. If genius is 1 per cent inspiration and 99 per cent perspiration, it was more heavily loaded towards inspiration in Teller's case. It was his open-mindedness, intellect and imagination that enabled him to make such an individualistic contribution to so many problems.

Edward Teller. (Photo by Bryan Quintance, by courtesy of the Lawrence Livermore National Laboratory)

He had discussed the possibility of obtaining energy from a thermonuclear device with Fermi, one day in the spring of 1942, when they were lunching together at Columbia University. The idea was that energy might be liberated by getting two very small atoms to join together, just as it was by splitting a large atom in two. The splitting had been called fission; the joining together was called fusion. The snag was that two small atoms would only fuse together at a very high temperature. It was known, for example, that the enormous energy of the sun was created by fusion reactions at the very high solar temperatures. The new idea of Fermi and Teller was that a fission bomb might be used to heat a mixture of small atoms to such a high temperature that they would fuse together and liberate energy. The atoms originally considered as most likely to undergo fusion at high temperature were hydrogen and its two isotopes, deuterium and tritium. Hydrogen is the smallest known atom, so it is allotted a relative atomic mass of 1 and symbolized as ^1H. The deuterium atom is twice as heavy, ^2H or D, and the tritium atom three times as heavy, ^3H or T. If what was being suggested could be achieved, there would be no limit to the size and power of bomb that could be built. More and more hydrogen and/or its isotopes could be packed around a central fission bomb, which would act simply as a giant detonator by raising the temperature to the required value.

This germ of an idea did not make any great impact on Fermi's mind, but Teller pondered it very thoroughly for a few weeks only to reach the conclusion that it was impracticable. In the early summer, however, he moved to the Metallurgical Laboratory at the University of Chicago and, finding himself at something of a loose end, he took the matter up again in collaboration with Emil Konopinski. The two of them quickly discovered that Teller's original conclusion was false. They decided that high temperature fusion of small atoms, though extremely difficult to achieve, might well be possible, and Teller became almost obsessed with the idea.

He had been one of the first to join the Los Alamos team, in April 1943, and he helped Oppenheimer with much of the early planning. The two were in fact quite friendly at that stage, but there was something in their natures which did not gel so that, although Teller was a great admirer of Oppenheimer professionally, they never got on well together. Nor was the relationship helped when Oppenheimer neither offered Teller a position as a leader of any of the major divisions at Los Alamos nor gave much support to the idea of aiming to build a fusion bomb alongside the fission bomb.

Any plans for actually building a fusion bomb were, in fact, shelved, but Teller was allowed to direct a small group to study the theoretical implications. This gave him a satisfying independence but it aroused a good deal of hostility when other scientists felt that he was not playing his full part in the main project, and Teller became a bit of an outsider. Nor did the group make much practical progress so that, by the time 'Little Boy' and 'Fat Man' were ready for use, they had no detailed ideas as to how a hydrogen bomb might be built though they confidently predicted that it would be possible to manufacture one with the power of 10,000,000 tonnes of TNT. What they had in mind consisted of a central fission bomb surrounded by about 1 cubic metre of liquid deuterium together with some tritium to act as a booster.

Fortunately they had discounted Teller's wildly alarming prediction that the explosion of a fission bomb might trigger off a fusion process in the hydrogen atoms contained in water vapour in the atmosphere causing a catastrophic, universal explosion which would blow up the whole world.

* * *

The great majority of the Los Alamos scientists were overjoyed when they heard that both 'Little Boy' at Hiroshima and 'Fat Man' at Nagasaki had exploded according to plan, and there were great celebrations. But the mood quickly changed when they began to look into the uncertain future and when details of the damage that the bombs had caused came through. For many it was a period of considerable soul-searching as they pondered on their own part and responsibility in making the bombs. What should, or could, be done in the future? Most scientists certainly wanted to get rid of the military control, represented by General Groves, and to be free to carry out their own research work once again. Many, led by Einstein, wanted to ban the bomb, and there were strong pleas for some sort of international control if there could be no ban.

No one agonized more than Oppenheimer because no one had been through the mill as much as he had. He had lost about 9 kg in weight during his time at Los Alamos and he had let Groves know, as early as May 1945, that he was contemplating resigning his post as soon as it was convenient. The actual resignation came on 16 October and he returned to university life, first at the California Institute of Technology and in 1947 as director of the Institute for Advanced Study at Princeton, New Jersey. It may well have been that he intended to divorce himself from nuclear energy matters altogether and he certainly told Teller that he would have nothing further to do with thermonuclear work. He also began to use emotive phrases such as 'The physicists have known sin' and 'We have blood on our hands.'[2] President Truman, it is said, retorted, 'Never mind. It'll all come out in the wash.' But with his unique experience and expertise, Oppenheimer was soon drawn into the maelstrom of the heated debate that was taking place between the scientists, the military and the politicians. When the McMahon Act was passed in the United States, in July 1946, it established civilian control over atomic matters through a new Atomic Energy Commission, consisting of five commissioners with equal power and equal voting rights. To provide it with technical information and judgements, a General Advisory Committee was set up in January 1947, and Oppenheimer was selected, by its nine members, as its first chairman. He held the post until 1952 so that he was, once again, at the very centre of things.

He was succeeded at Los Alamos by Norris Bradbury but, as so many of the workers there followed Oppenheimer back into civilian life, it was a struggle to keep things going and morale was low. Enough progress was made, however, to enable the successful completion of Operation 'Crossroads' off Bikini atoll in July 1946, which involved a task force of 40,000 men. Two atom bombs were exploded, one above and one below water, to see what impact they had on ships, so that plans for future naval strategy and design could be drawn up.

Edward Teller did not at first join the general exodus from Los Alamos. Some who did not like him even suggested that he hung around hoping to get Oppenheimer's job. In the event Bradbury offered him a post as head of the Theoretical Division – a position he had rather hoped to occupy in the Oppenheimer era – but Teller turned it down when he discovered that the sort of expansive programme which he envisaged would not be possible because of lack of resources. So he followed Fermi to the University of Chicago where he lectured on quantum mechanics, did some research, and appeared to be settling down to enjoy a more relaxed lifestyle. But he still pursued his interest in thermonuclear matters and made frequent visits to Los Alamos to keep in touch with developments there.

On one such visit, in April 1946, he took the chair at a secret meeting to review the progress being made on the super-bomb project. Almost all those in any way involved were present, including Klaus Fuchs, and there was general agreement that a super-bomb could be constructed but some difference of opinion about the time-scale involved. The project would use up a high proportion of the resources available within the atomic energy field, and a decision as to the best way to move forward could only be taken as part of the highest national policy. But in 1946 national policy was not in favour of the super-bomb, even though Teller was still rooting for it and predicting that it could be built in two years.

* * *

It was not until September 1949, when it was discovered by aerial surveillance that the Russians had exploded their first atom bomb, that events began to move in Teller's direction. The news came as a great surprise and shock to the United States, where it had been assumed that it would be at least 1956 before Russia would be able to make such a bomb. Teller's view was that an arms race was no longer a possibility but a frightening certainty which had to be faced. For the United States to keep ahead, the Russian bomb, nicknamed Joe I after Joseph Stalin, must be trumped by a super-bomb. So, just as the threat of a German atom bomb had lent so much urgency to the work in Great Britain and the United States on a fission bomb, so the reality of a Russian fission bomb and the threat of a Russian super-bomb spurred on Teller and his supporters.

The General Advisory Committee, with Oppenheimer in the chair, met on 19 October 1949 to consider whether to recommend the production of a thermonuclear bomb in response to the news from Russia. All eight members present at the meeting turned the idea down. They agreed that its manufacture was technically feasible but that it would be so difficult and so costly that it would seriously interfere with the smooth development of the ever-expanding programme for building fission bombs. They pointed out that in their judgement there was no need for such a powerful bomb and that it would do nothing to improve the defences of the United States. But most importantly they stressed that the moral position of the country would be damaged if it decided to make such a vast leap forward in the arms race.

Teller and his supporters were flabbergasted by the attitude of the General Advisory Committee but it only encouraged them to step up their lobbying in favour of the super-bomb. And before long they had General Bradley, the Chairman of the Joint Chiefs of Staff, Louis Johnson, the Secretary of Defense, and Dean Acheson, the Secretary of State, on their side. Once again Truman was confronted with a crucial decision and his advisers were split. The scientists were saying 'No' and the military were saying 'Yes'.

The issue must have been balanced on a knife-edge and the scales may very well have been tipped by the announcement on 27 January 1950 that Klaus Fuchs had been arrested in London after voluntarily confessing to the police about his spying activities over the previous eight years. He had been born in Germany in 1911 and he had joined the Communist party in that country as a 21-year-old student activist, mainly because his family had suffered so much under the Nazis. His Marxist philosophy gave him a 'complete confidence in Russian policy' but he had to flee from the Nazis, and he arrived in England in 1933 virtually penniless and friendless. After working in the universities of Bristol and Edinburgh, and after a period of internment in Canada at the start of the war, he was invited by Peierls to work with him on the separation of uranium isotopes at Birmingham University, and to live with him and his family. The British Security

Services were still suspicious of Fuchs but police enquiries could find no evidence against him since his arrival in Great Britain so he was granted a full clearance and was, eventually, naturalized. Peierls took Fuchs with him to New York and Los Alamos, and neither he nor his wife could bring themselves to believe the news that he was a traitor until they heard it from his own mouth when visiting him in Brixton Prison. Lady Peierls was extremely angry and wrote to him saying, 'You have burned your God. God help you.' Fuchs attributed his activities to a 'controlled schizophrenia' but that did not prevent him from being sentenced to fourteen years' imprisonment and being stripped of his British citizenship behind which he had sheltered. He was a model prisoner so he gained the normal maximum remission of sentence and was released after nine years, when he went to East Germany to work at the Institute for Nuclear Research at Rossendorf. By a remarkable quirk of fate, he found himself in charge of the establishment when the existing director, Dr Barwick, defected to the West. Fuchs was awarded the Order of Karl Marx in 1981 and died on 28 January 1988.

It is now clear that it was the information passed to the Russians by Fuchs that had enabled them to build a fission bomb so rapidly, and it was also probable that he had provided them with much useful information about fusion bombs because Fuchs had been working on them at Los Alamos. Knowing that, it took Truman only four days to decide between his scientific and his military advisers. He issued his verdict on 31 January, declaring:

> It is part of my responsibility as Commander-in-Chief of the Armed Forces to see that our country is able to defend itself against any possible aggressor. Accordingly, I have directed the Atomic Energy Commission to continue its work on all forms of atomic weapons, including the 'hydrogen' or super-bomb. Like all other work in the field of atomic weapons, it is being and will be carried forward on a basis consistent with the over-all objectives of our programme for peace and security.[3]

* * *

Teller must have thought that building a super-bomb, now that the go-ahead had been given, would be simple compared with all the in-fighting of the previous months and years, but he was in for a rude shock. He had by chance taken time off from Chicago to travel abroad and to return to work at Los Alamos where Bradbury was reorganizing the programme to make way for a maximum effort on fusion without affecting too much the continuing work on fission. Alas, Bradbury could get on with Teller no better than Oppenheimer had and he judged that he might have a mutiny on his hands if he put Teller in charge of anything very important, but he did appoint him as assistant director of weapons development and made him chairman of a committee which kept an eye on thermonuclear progress.

The maximum effort on fusion got off to a very disappointing start and progress throughout most of 1950 was backwards. Stanislaw Ulam, a refugee from Poland who had left the University of Wisconsin to work at Los Alamos in the winter of 1943, had carried out a mathematical reassessment of Teller's idea of using a mixture of deuterium and tritium together with a fission bomb to make a super-bomb. Working with only one assistant and with simple slide-rules, Ulam discovered that Teller's idea simply would not work. Teller was furious and near to despair at the thought of all his previous efforts having been in vain. He could not, at first, bring himself to regard Ulam's calculations as valid but, when his results were confirmed by an analysis carried out on one of the first electronic computers, nicknamed ENIAC (Electronic Numerical Integrator And Calculator), he had to reconsider the situation and go back to the drawing-board. Ulam's contribution was so important that it has been suggested that he should be regarded as the father of the hydrogen bomb – with Teller as the mother, because he had held the baby for so long.

In simple terms Teller's original design for a super-bomb, in which a deuterium-tritium mixture was packed around a central fission bomb, would not work because the explosion of the fission bomb would blow the deuterium-tritium mixture apart before it had been heated to a sufficiently high temperature for the atoms to fuse together.

That might have been the end of the story had not Ulam and Teller, though still smarting from their earlier disagreement, come up with a new solution in the spring of 1951. They realized that there are two surges or bursts of energy when a fission bomb explodes, one causing the heat effect and the other the blast effect. The heat effect comes from an intense concentration of X-rays, which moves at the speed of light, while the blast effect is caused by a slower-moving shock wave. It was a matter of ensuring that the X-rays were able to raise the deuterium-tritium mixture to the required temperature before the shock wave could disrupt it. To achieve this, the fission bomb and the deuterium-tritium mixture had to be some distance apart, but within the same container, so that the fast-moving X-rays reached the mixture a fraction of a second before the slow-moving shock wave. The mixture would then be heated up by the X-rays before it could be blown apart by the shock wave.

The effect of the X-rays on the deuterium-tritium mixture could be further enhanced by surrounding the mixture by a dense plastic material. The impact of the X-rays on this material would produce a region of both very high temperature and very high pressure, so that the deuterium-tritium mixture at the centre of the plastic would not only be heated but also compressed. And the fact that it was being compressed, with the atoms being pushed closer together, meant that it would fuse at a lower temperature.

Moreover, it was realized that the pressure created within the plastic material was high enough to compress a piece of plutonium sufficiently for it to become supercritical in size as happened, under pressure, in 'Fat Man'. As a refinement, then, it would be possible to make a still more powerful bomb by enclosing some plutonium within the deuterium-tritium mixture. The original fission bomb would cause fusion of the deuterium-tritium mixture and at the same time a further fission of plutonium.

These new ideas were outlined by Teller at a high-level, weekend conference organized and chaired by Oppenheimer at Princeton in July 1951. Almost everyone concerned with the thermonuclear programme – the Atomic Energy Commissioners, military representatives, Los Alamos scientists and independent consultants – were present, and there was an unusual measure of agreement at the end of the meeting that 'we had something for the first time that looked feasible in the way of an idea'. Oppenheimer described the idea as 'technically sweet'[4] and recommended that it be pursued with none of the reservations that had been expressed so plainly in the report of the General Advisory Committee in the previous year. Later, he was to explain the discrepancy on the grounds that the tone of the earlier report would have been different if the General Advisory Committee had known in 1949 what they knew in 1951.

The success of this Princeton meeting gave the super-bomb project a big push forward. And it was not overdue because eighteen months had passed since Truman had directed that the bomb should be built. The working week at Los Alamos was extended from five to six days, and those who operated ENIAC did night and day shifts. The calculations were, however, so detailed and complicated that it was not until a much more powerful computer, humorously nicknamed MANIAC (Mathematical Analyser, Numerical Integrator And Calculator), was designed that the problems were solved.

A series of four preliminary tests, code-named 'Greenhouse', had been carried out at Eniwetok in May 1951. They had been designed to show beyond doubt that a mixture of deuterium and tritium would explode, via a fusion reaction, if heated to a high enough temperature. It was the detection of high energy neutrons, as a by-product, that demonstrated that a fusion reaction had taken place. These were basic, scientific tests, and not tests on anything that could be called a bomb, but the explosive power available was clear for all to see when two small islands of the atoll – Eberiru and Engebi – were completely blown up.

'Mike' was the first self-contained thermonuclear device to be tested by the Americans. It could not really be regarded as a deliverable bomb as it weighed 66 tonnes, mainly because of all the refrigeration equipment necessary to keep the deuterium-tritium mixture below −253°C so that it remained liquefied. To cut down the weight it was necessary to replace the deuterium-tritium mixture by a solid material in what came to be known as a dry bomb. The compound between a lithium-6 isotope, ^{6}Li, and deuterium, lithium deuteride, was found to be satisfactory. Bombardment of this solid, within a bomb, by neutrons, converts some of the ^{6}Li atoms into tritium and, as deuterium atoms are already present, a satisfactory 'fuel' mixture for fusion is created.

The first dry bomb, code-named 'Shrimp', was tested at Bikini in Operation 'Castle', in the spring of 1954. It proved to be much more powerful than had been expected, with a blast equivalent to 15,000,000 tonnes (15 megatonnes) of TNT; it left a crater 182 m wide and 68 m deep. As it only weighed 10,645 kg it could be carried in the bomb bay of a B-47, so it was the first ever practical hydrogen bomb. A series of further tests on devices such as 'Runt', 'Koon', 'Yankee' and 'Nectar' confirmed that lithium deuteride was, indeed, a satisfactory fuel for hydrogen bombs and that they could be made in a variety of sizes. America had taken the lead but others soon joined the bandwagon. Russia exploded their first real hydrogen bomb in November 1955, though they had tested a preliminary

The second test of the British H-bomb on 31 May 1957. The bomb was dropped from a V-bomber high over the central Pacific. (Imperial War Museum, GOV9121)

bomb (Joe 4) as early as August 1953; the United Kingdom had its H-bomb by May 1957; China by June 1967; and France by August 1968.

The original international arms race, in the 1940s, had been between Nazi Germany on the one side and America, Great Britain and France on the other. Now, despite many international attempts, generally initiated by those countries with existing nuclear weapons, to limit what came to be known as nuclear proliferation, it was expanding on a wider scale. The race spread into Southern Asia when India, troubled by events in China, tested a nuclear device underground in 1974. She described it as a 'peaceful nuclear explosion' but it was not viewed like that by Pakistan, which set to work to build up its own nuclear capability. Other countries, notably Israel, South Africa, Brazil, Iran, Iraq, North Korea and Argentina, reacted in the same way, and there are today many more countries with the necessary technical ability to move in that direction if they ever felt the need. First India, and then Pakistan, caused much alarm by each carrying out a series of underground nuclear tests in May 1998.

*   *   *

Teller had been present at the 'Greenhouse' tests but he was not there for the testing of 'Mike' because he had resigned from his post at Los Alamos in July 1952. His relationship with Bradbury, and with others of the Los Alamos team, had deteriorated during the time he had been working there until he began to feel that satisfactory progress towards his goal would only be made if he could himself control a laboratory devoted entirely to thermonuclear work. He gained support for this idea, particularly from the Air Force. Neither the Atomic Energy Commission nor the General Advisory Committee were originally in favour of Teller's idea but eventually a small laboratory at Livermore, belonging to the University of California, was taken over. It was staffed with bright young men, many of them with little or no experience in the field of atomic energy, and, at first their bomb designs were flops but, as they gained in experience, they contributed much to the development of nuclear weapons.

Bradbury had in fact formally invited Teller to attend the 'Mike' test, but the invitation had, understandably, been declined. Teller nevertheless felt that he wanted to keep in touch with the test and he had the idea of doing so via a seismograph in the basement of the Geology Department at the University of California. He calculated that the shock waves from the explosion would take about 15 minutes to travel the thousands of kilometres under the Pacific from Eniwetok to the Californian coast. And so, in due course, they did, making 'clear and big and unmistakable' impressions on the photographic plate from the seismograph. So Teller knew that the test had been a success almost as soon as those who were present.

With the success of the H-bomb tests, Teller's star was very much in the ascendant, while Oppenheimer's had waned. Relationships between the government and the General Advisory Committee had not been very cordial since Truman had overridden the Committee's recommendation in 1950, and Oppenheimer resigned from the chairmanship of the Committee in July 1952. He was still engaged as a consultant to the Atomic Energy Commission but he was consulted less and less.

He had been regarded by some as a security risk since before his appointment as director at Los Alamos, and he had been kept under constant surveillance and frequently interrogated. He said that the government must have spent much more on tapping his phone than they ever paid him, and J. Edgar Hoover, the head of the Federal Bureau of Investigation, had testified against him at a security review in 1948. Though Oppenheimer had survived that, there were still many behind the scenes who were highly suspicious of him.

Things came to a climax in 1953. Oppenheimer had spent the summer lecturing in South America and had been to England in November to deliver the Reith Lectures for the British Broadcasting Corporation and to receive an honorary degree at Oxford University. He then visited various friends in Europe including Haakon Chevalier in Paris. While he was away from the United States, he was constantly shadowed, and William Borden, a young executive who was investigating charges of mismanagement in the Atomic Energy Commission on behalf of a Congressional Committee, was finishing a lengthy document, which he sent to the director of the FBI, arguing that Oppenheimer was probably a Soviet agent.

When Oppenheimer returned home he was summoned to a meeting in Washington with Lewis L. Strauss, the Chairman of the Atomic Energy Commission, who had been directed by President Eisenhower to look into the allegations. It was perhaps an unfortunate choice because Strauss, who had left school at sixteen to sell shoes but eventually became a millionaire financier and a rear-admiral in the Naval Reserve, had little in common with Oppenheimer and rather openly despised him. He was also very self-confident, somewhat pompous, and could be obstinate and manipulative. Their first meeting took place on 21 December, and Oppenheimer was staggered to learn that his security clearance was to be removed because of doubts about his 'veracity, conduct and even loyalty'. The gist of the charges against him related to his association over the years with Communists or their sympathizers but there was an added, surprising claim that he had strongly opposed the development of the H-bomb project in the postwar years.

Oppenheimer was told that he could either resign without any further investigation of the charges laid against him or he could have the charges referred to a Personnel Security Board selected by the Atomic Energy Commission. He refused to resign, writing to Admiral Strauss:

> I have thought most earnestly of the alternative suggested. Under the circumstances this course of action would mean that I accept and concur in the view that I am not fit to serve this government, that I have now served for some twelve years. This I cannot do. If I were thus unworthy I could hardly have served our country as I have tried or been the director of our Institute in Princeton or have spoken, as on more than one occasion I have found myself speaking, in the name of science and our country.[5]

The hearing against Oppenheimer began in private on 12 April 1954, before a three-man board, and lasted for four weeks. It was not a legal trial but the procedure adopted was very much like that of a court of law, with cross-examining of witnesses by professional lawyers. Roger Robb, representing the Atomic Energy Commission, behaved like a robust, merciless public prosecutor, while Oppenheimer was defended by Lloyd Garrison, a mild-mannered, scholarly attorney, who had devoted himself to many worthy public causes but who had little courtroom experience.

About forty eminent scientists, politicians and servicemen gave evidence at the hearing, with the majority speaking in favour of Oppenheimer. Among the scientists, Bradbury supported him particularly strongly and every member of the General Advisory Committee was prepared to testify on his behalf. Teller also said that 'Oppenheimer had made just a most wonderful and excellent director of the Los Alamos Laboratory'[6] but some of his other testimony was very damaging. In reply to one question from Robb, enquiring about Oppenheimer's loyalty, he said: 'I know Oppenheimer as an intellectually most alert and complicated person, and I think it would be presumptuous and wrong on my part if I would try to analyse his motives. But I have always assumed, and I now assume, that he is loyal, to the United States. I believe this, and I shall believe it until I see very conclusive proof to the opposite.'[7] But when asked for his views on Oppenheimer as a security risk he replied:

> In a great number of cases, I have seen Dr Oppenheimer act – understood that Dr Oppenheimer acted – in a way which for me was exceedingly hard to understand, and his actions frankly appeared to me confused and complicated. To this extent I feel I would like to see the vital interests of this country in hands which I understand better and therefore trust more. In this very limited sense I would like to express a feeling that I would feel personally more secure if public matters would rest in other hands.[8]

Later, in a cross-examination lasting for more than an hour, he said: 'If it is a question of wisdom and judgement, as demonstrated by actions since 1945, then I would say one would be wiser not to grant security clearance.'[9] Such damning evidence so upset many of his fellow scientists that Teller was ostracized for many years.

Yet it was Oppenheimer's own showing in the courtroom that was most damaging to his cause. He was known as a brilliant speaker, able, with great eloquence, to present almost any case, but when it came to speaking on his own behalf he was inarticulate and extremely diffident and apologetic. It was as though he had been struck dumb. Robb attacked him relentlessly, treated him rather like a common criminal, and lured him into many pitfalls. He forced him to admit, for example, that he had lied, ten years previously, when relating his version of his association with Haakon Chevalier and, when asked by Robb why he should have done that, he replied, 'Because I was an idiot.'[10] In the evening, after that hearing, Robb told his wife that he had 'just seen a man destroy himself'.[11]

The most important conclusion of the board, by a majority of two to one, was that Oppenheimer was 'a loyal citizen' but that they could not recommend a continuation of his security clearance. The majority view was that Oppenheimer's 'continuing conduct and associations have reflected a serious disregard for the requirements of the security system'; that he was 'susceptible to influences which could have serious implications for the security interests of the country'; that 'his conduct in the hydrogen bomb programme was sufficiently disturbing as to raise a doubt as to whether his future participation, if characterized by the same attitudes in a government programme relating to the national defence, would be clearly consistent with the best interests of security'; and that 'he had been less than candid in several instances in his testimony before the Board'.

Oppenheimer appealed against the judgment to the Atomic Energy Commissioners, but they rejected the appeal by four votes to one. The majority did not express any clear opinion on Oppenheimer's loyalty, but they referred to 'fundamental defects in his character', regarded his associations with 'persons known to him to be Communists' as being 'extended far beyond the tolerable limits of prudence and self-restraint which are to be expected of one holding the high positions that the government had continuously entrusted to him since 1942', and concluded that he had 'fallen far short of the acceptable standards of reliability, self-discipline and trustworthiness'[12] to which a government official must measure up.

So a much chastened Oppenheimer went back to directing the activities of the Institute for Advanced Study at Princeton, despondent and 'eating out his heart in frustration'.[13] But time is a great healer and, as the years passed, more and more people came to think that he had been shoddily treated and some attempt at official recompense was made when President Kennedy agreed in 1963 to carry out a recommendation by the Atomic Energy Commissioners that he should present Oppenheimer with the Enrico Fermi Award. This prize, consisting of a citation, a gold medal and the sum of 50,000 dollars, had been created by the Atomic Energy Commission in 1955 to honour those who had made 'especially meritorious contributions to the development, use, or control of atomic energy'. Sadly, President Kennedy was assassinated before he could present the award, but it was given to Oppenheimer by President Johnson on 2 December 1963, exactly twenty-one years, to the day, since Fermi's CP-1 had operated successfully in the squash court in Chicago.

Teller, too, slowly came to be accepted, at least partially, back into the scientific fold. He had been presented with the Enrico Fermi Award in 1962 and he was present at its award to Oppenheimer a year later. As the two came face to face there was an embarrassing pause before they shook hands.

* * *

The Atomic Energy Commission was not only responsible for the removal of Oppenheimer from the atomic energy scene; its setting up had earlier led to the departure of General Groves. He was very ill-pleased by the transfer of the vast empire, of which he was the unchallenged tsar, to civilian control and, as a member of the Military Liaison Committee, supposedly advising the Commission, he spent much of his time trying to insert spokes into the wheels of the new organization. He seemed to harbour the hope that civilian control would not work successfully and that he would have to be recalled to sort out the mess. But that was not to be, and, after an acrimonious fight over a period of about two years the general was removed from the Liaison Committee in September 1947. Shortly after, his long and brilliantly effective association with atomic energy came to an end when he resigned from the Army to become a vice-president, in charge of research, of the Remington Rand Corporation. He retired fully in 1961 and died on 13 July 1970.

*   *   *

Lewis Strauss had been a member of the Atomic Energy Commission from 1947 to 1950, and was its chairman between 1953 and 1958. At the end of that time he was invited by President Eisenhower to serve a second term as chairman but he resigned because he thought that the Senate might block his re-election. A year later, Eisenhower nominated him as the Secretary of State for Commerce but, after a two-month gruelling Senate hearing in which he was 'pilloried for his arrogance and rigidity, and caught in a lie under oath',[14] confirmation of his appointment was refused and Strauss, despondent and humiliated, retired from public life. He had, at last, found something in common with Oppenheimer.

*   *   *

Let Oppenheimer, the brains behind Groves's bravura, and a giant in this thousand-year-long tale of explosives, have the last word. On the day of his retirement from Los Alamos in 1945, he accepted, on behalf of the laboratory, a Certificate of Appreciation from the Secretary of State for War which was presented by General Groves. Oppenheimer responded:

> It is with appreciation and gratitude that I accept from you this scroll for the Los Alamos laboratory, for the men and women whose work and whose hearts have made it. It is our hope that in years to come we may look at this scroll, and all that it signifies, with pride.
>
> Today that pride must be tempered with profound concern. If atomic bombs are to be added as new weapons to the arsenals of the warring world, or to the arsenals of nations preparing for war, then the time will come when mankind will curse the names of Los Alamos and Hiroshima.
>
> The peoples of this world must unite or they will perish. This war that has ravaged so much of the earth has written these words. The atomic bomb has spelled them out for all men to understand. Other men have spoken them, in other times, of other wars, of other weapons. They have not prevailed. They are misled by a false sense of human history who hold that they will not prevail today. It is not for us to believe that. By our works we are committed to a world united, before this common peril, in law, and in humanity.[15]

After a life of massive achievement, coupled with much torment, he died of cancer of the throat at Princeton on 18 February 1967, at the age of sixty-two. Shortly before he died he had written: 'Science is not everything, but science is very beautiful.'[16] He now rests in peace. But the world, despite recent attempts to achieve nuclear disarmament, is still in some turmoil and he would doubtless think that it had done too little to heed his 1945 message of hope.

APPENDIX I

Names and Formulae

Historically, elements and compounds were named as they were discovered or made and, in the circumstances, the system that developed worked reasonably well. So much so that many of the old, traditional names are, in fact, still used, particularly in everyday life. Today, however, rules covering the nomenclature of chemicals, worked out under the auspices of the International Union of Pure and Applied Chemistry (IUPAC), have been widely adopted. They aim to give each individual chemical its own unique name which is related to its structure and to that of similar substances. Because there are so many different chemicals this is a daunting task, not unlike that of trying to name every person in the world differently.

The two main sets of rules cover all organic and inorganic compounds. The former are compounds associated with living matter, containing, mainly, carbon, hydrogen, nitrogen and oxygen atoms. The latter, not generally linked with living matter, are made up from all the other atoms. They include, for example, the many minerals which contain metallic atoms.

The different states of a substance are represented by (s) for solid, (l) for liquid or (g) for gas or vapour after its name. Thus ice is $H_2O(s)$, water $H_2O(l)$ and steam $H_2O(g)$.

CHEMICAL REACTIONS

A chemical reaction involves a rearrangement of atoms. The bonds in the molecules of the reagents split and the free atoms released, which exist only momentarily, recombine to give the different molecules of the products. The reaction is commonly summarised in the form of a chemical equation.

There are three main requirements for the reaction to be explosive. First, it must take place very quickly. Secondly, it must be an exothermic reaction; that is, heat must be evolved. This arises when more energy is given out in forming the bonds in the molecules of the products than is required to break the bonds in the molecules of the reagents. Thirdly, as many of the products as possible must be gases. As they will be hot, this ensures a big rise in pressure which is the main cause of the explosion.

GUNPOWDER AND ITS MODIFICATIONS

Gunpowder is made from carbon (C), sulphur (S) and potassium nitrate (KNO_3). No single chemical equation can fully represent the nature of the complex chemical reaction that occurs when it explodes but an over-simplification commonly used is as follows

$$4KNO_3(s) + 7C(s) + S(s)$$
$$\downarrow$$
$$3CO_2(g) + 3CO(g) + 2N_2(g) + K_2CO_3(s) + K_2S(s)$$

Less than half the gunpowder is converted into gaseous products and much of the heat produced in the explosion is retained in the solid products. Both these factors limit the efficiency of gunpowder as an explosive.

ANFOs, made from ammonium nitrate and fuel oils, are more efficient. The equation for the reaction of a typical example,

$$25NH_4NO_3(s) + C_8H_{18}(l) \qquad 8CO_2(g) + 59H_2O(g) + 25N_2(g)$$

shows that all the explosive mixture is converted into gases.

Other chemicals used in making gunpowder substitutes include

$NaNO_3$	NH_4NO_3	$KClO_3$
Sodium nitrate	Ammonium nitrate	Potassium chlorate(V)

$NaClO_3$	$KClO_4$	NH_4ClO_4
Sodium chlorate(V)	Potassium perchlorate (chlorate(VII))	Ammonium perchlorate (chlorate(VII))

EXPLOSIVES CONTAINING THE NITRO $-NO_2$ GROUP

A number of explosives are made by replacing the hydrogen atoms within the C–H or N–H bonds in organic compounds by nitro $-NO_2$ groups by treatment with a mixture of nitric (HNO_3) and sulphuric (H_2SO_4) acids. The process is known as nitration and is summarised as

$$-\overset{|}{\underset{|}{C}}-H + H-O-NO_2 \xrightarrow{-H_2O} -\overset{|}{\underset{|}{C}}-NO_2$$

$$-\overset{|}{N}-H + H-O-NO_2 \xrightarrow{-H_2O} -\overset{|}{N}-NO_2$$

If only one hydrogen atom in a molecule, X, is replaced by a nitro group the product is called nitro–X or mononitro–X; if two hydrogen atoms are replaced, it is dinitro–X; if three, tri-nitro–X. The precise position of the nitro group within a molecule is shown by the use of numbers.

Trinitrotoluene, TNT. Toluene got its name because it was made, in 1841, by distilling Tolu balsam and because it was like benzene. Its modern name is methyl benzene and it is converted into TNT by nitration.

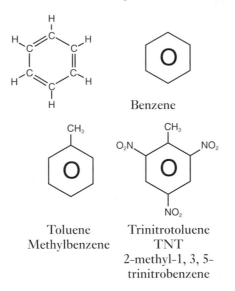

Benzene

Toluene
Methylbenzene

Trinitrotoluene
TNT
2-methyl-1, 3, 5-
trinitrobenzene

When TNT explodes, its molecules split up, partially because the C–N bond is weak; the equation is

$$C_7H_5N_3O_6(s) \rightarrow 3.5CO_2(g) + 2.5H_2O(g) + 1.5N_2(g) + 3.5C(s)$$

Because there are too few atoms of oxygen within the molecule some of the carbon remains as a solid and is not oxidised into gases. That is why black smoke is produced in a TNT explosion and why ammonium nitrate is added in the amatols to provide more oxygen.

Lyddite. Lyddite is made by nitrating phenol. Its old chemical name was picric acid (from the Greek *pikros* = bitter).

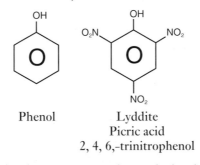

Phenol

Lyddite
Picric acid
2, 4, 6,-trinitrophenol

The simple equation that best represents its explosion is

$$C_6H_3N_3O_7(s) \rightarrow 5.5CO(g) + 1.5H_2O(g) + 1.5N_2(g) + 0.5C(s)$$

RDX. RDX or cyclotrimethylenetrinitramine is made by the nitration of hexamine or hexamethylene tetramine.

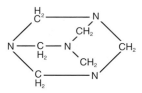

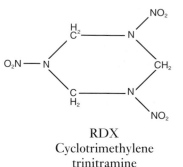

Hexamine
Hexamethylene
tetramine

RDX
Cyclotrimethylene
trinitramine

The equation for its explosion is

$$C_3H_6N_6O_6(s) \rightarrow 3CO(g) + 3H_2O(g) + 3N_2(g)$$

HMX and HNIW.

HMX
Cyclotetramethylene-
tetranitramine

HNIW
Hexanitrohexaza-
isowurtzitane

EXPLOSIVES CONTAINING THE NITRATE –O–NO₂ GROUP

Nitrates are formed by reaction between a mixture of nitric and sulphuric acids and organic compounds containing hydroxyl, –OH groups. In summary,

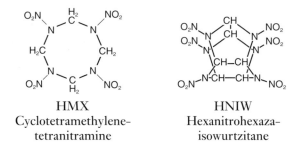

Many of the products have been and still are wrongly regarded as nitro-compounds. They do contain an –NO₂ group but it is linked to an oxygen atom so that it is, really, just a part of a nitrate group.

Dynamite. The main component of dynamite is still commonly called nitroglycerine or nitroglycerol but the more correct name is glycerol trinitrate or, better still, propane–1,2,3–triyl trinitrate.

$$
\begin{array}{cc}
\text{H}_2\text{C–OH} & \text{H}_2\text{C–O–NO}_2 \\
\text{HC–OH} & \text{HC–O–NO}_2 \\
\text{H}_2\text{C–OH} & \text{H}_2\text{C–O–NO}_2
\end{array}
$$

Glycerine   Nitroglycerine

Glycerol   Nitroglycerol

Propane–1,2,3–triol   Glycerol trinitrate

Propane–1,2,3–triyl trinitrate

The equation for its explosion is

$$\text{C}_3\text{H}_5\text{N}_3\text{O}_9(\text{l}) \rightarrow 3\text{CO}_2(\text{g}) + 2.5\text{H}_2\text{O}(\text{g}) + 1.5\text{N}_2(\text{g}) + 0.25\text{O}_2(\text{g})$$

Guncotton. When cellulose is treated with a mixture of nitric and sulphuric acids various products are formed. Prolonged treatment with hot, concentrated acids gives a product which contains about 13.3 per cent of nitrogen and is called guncotton; its main component is cellulose trinitrate. Weaker acids at a lower temperature for a shorter time give a mixture containing between 8 and 12 per cent of nitrogen. It is known as pyroxylin or collodion and is a mixture of cellulose mono- and di-nitrates.

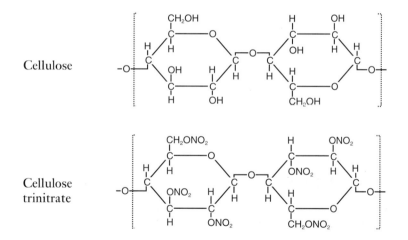

PETN. PETN or pentaerythritol tetranitrate is made by treating penta-erythritol with a mixture of nitric and sulphuric acids.

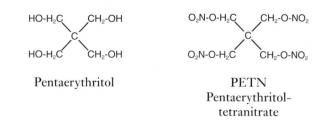

Pentaerythritol   PETN

Pentaerythritol-

tetranitrate

INITIATING EXPLOSIVES

Initiating explosives are particularly sensitive. Some typical examples are given below:

$Hg(CNO)_2$ $Pb(N_3)_2$ $(NO_2)_3C_6HO_2Pb$

Mercury fulminate   Lead azide   Lead styphnate
Lead 2,4,6–trinitro-
resorcinate

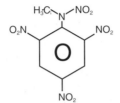

Tetryl (CE)
2,4,6–Trinitrophenyl-
methylnitramine

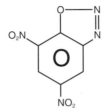

DDNP. DINOL
Diazodinitrophenol

POSSIBLE NEW PRODUCTS

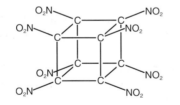

Octanitrocubane

Octa-azacubane

APPENDIX II

Energy and Power

Explosions occur when there is a rapid release of energy which produces a large volume of hot gases and a build up of pressure. In a chemical explosion, the energy is released in a chemical reaction; in a nulear explosion, it is nuclear energy that is released. The SI unit of energy is the joule, with 1 J being the work required to move a force of 1 newton over a distance of 1 metre. That is equivalent to lifting a small apple, of mass 102 g, by 1 metre. 1 kilojoule (1 kJ) is 1000J and 1 Megajoule (1MJ) is 1000000J.

In both chemical and nuclear explosions the release of energy arises from the conversion of a small amount of mass into energy in the course of the explosion. As a conseqence of his theory of relativity, Albert Einstein put forward the idea, in 1907, that mass and energy were related according to the equation

$$E = m \times c^2$$

E	=	m	x	c^2
Energy	=	Mass	x	(Velocity of light)^2
(in joules)		(in kilograms)		(in metres per second)

Because the velocity of light is a very large number (2.997 924 580 x 10^8 m per sec) it follows that small amounts of mass can provide large amounts of energy. 1 kg of mass, indeed, is equivalent to 8.988 x 10^{16} joules.

Sir James Jeans referred to mass as 'bottled energy' but it is not easy to get complete or efficient conversion of mass into energy. The energy liberated in any ordinary chemical reaction requires a conversion of only about 100 x 10^{-12} kg of mass and this cannot be detected on any balance.

POWER

Energy is a measure of the capacity to do work whereas power is a measure of the rate at which the work is done. Thus, a man uses energy and does work when he lifts a suitcase. If he lifts it twice as quickly he uses the same energy but needs twice the power. The unit of power is the watt with 1 watt equivalent to 1 joule per second. The older unit of horse power, based by James Watt on an estimate that an average horse could lift a weight of 51 kg to a height of 60 m in 1 minute and could go on doing it for a whole shift, is now taken as 745.70 watt.

EXOTHERMIC REACTIONS

A chemical reaction which releases energy is called an exothermic reaction. The burning of hydrogen in air is typical; it is represented by the following equations,

$$2H_2(g) + O_2(g) \rightarrow 2H_2O(g) + 484kJ$$

$$2H_2(g) + O_2(g) \rightarrow 2H_2O(g) \quad \Delta H = -484kJ$$

The first equation indicates that 484 kJ of energy are released when 2 mol, 4 gram, or $12.044\ 273 \times 10^{23}$ molecules of hydrogen are completely burnt. The second equation provides the same information in a slightly different way by giving what is known as the enthalpy change, ΔH, for the reaction. It is the *change* in energy that has taken place. Beause energy has been released, the products of the reaction have a lower enthalpy than the initial reagents. That is why the ΔH value is negative.

In an endothermic reaction, energy is absorbed and the ΔH value is positive. For example,

$$N_2(g) + O_2(g) \rightarrow 2NO(g) - 90.4\ kJ$$

$$N_2(g) + O_2(g) \rightarrow 2NO(g) \quad \Delta H = 90.4\ kJ$$

In a simple way, an exothermic reaction can be regarded as a 'downhill' reaction; an endothermic one as 'uphill'.

Heats of explosion. The amount of heat released by an explosive is commonly known as the heat of explosion. It is usually expressed in joules per gram. Some typical, calculated values, are listed below;

Nitroglycerine	6275	PETN	5940
RDX	5130	TNT	4080

The experimentally measured values may be slightly different depending on the way the explosion is brought about.

The heat of explosion is calculated from bond energy values which give a measure of the amount of energy involved in breaking or forming a particular bond. The values are generally expressed in kilojoules (kJ) per mole of the bond, i.e. for 6.06×10^{23} individual bonds. Some typical values are

H–H	436	C–H	412	C–C	348
O=O	496	O–H	463	C=C	743
N≡N	944	N–H	388	C≡N	305

The burning or explosion of hydrogen and oxygen, according to the equation given above, requires (2 x 436) kJ to split two H–H bonds and 496 kJ to split one

O=O bond; that is 1368 kJ. Forming four O–H bonds releases (4 x 463), i.e. 1852 kJ. The amount of energy released is, therefore, (1852 – 1368), i.e. 484 kJ.

One of the reasons why chemicals containing C–N bonds are likely to be explosive is that there is a large release of energy when the weak C–N bond is converted into the much stronger N≡N bond.

THE RATE OF REACTION

An explosive depends more on the rate at which its energy is released than on the total amount of energy. 1 kg of TNT, for example will release 4080 kJ of energy when it is detonated. 1 kg of petrol will release more than 30,000 kJ when it is fully burnt in air, but because that burning is a relatively slow process the petrol will not normally explode. It can, however, be made to explode if it is well mixed with a lot of air before ignition so that the burning takes place much more rapidly.

Similarly, natural gas or methane will normally burn steadily in a gas cooker or fire. A mixture of a lot of air with only a little natural gas can, however, be explosive. It is the build up of such mixtures (fire damp) in coal mines that is the cause of many disasters.

NUCLEAR EXPLOSIONS

The energy release in a nuclear explosion comes about because different atoms have different bonding energies within their nuclei. As with a chemical explosion, it is the associated very slight loss of mass which produces the energy.

A possible way in which the nucleus of U might be split by a neutron ($_{0}^{1}$n) is summarised in the following equation

$$^{235}_{92}\text{U} \quad + \quad ^{1}_{0}\text{n} \quad \longrightarrow \quad ^{144}_{52}\text{Ba} \quad + \quad ^{90}_{36}\text{Kr} \quad + \quad 2^{1}_{0}\text{n}$$

235.0439	1.0087	143.881	89.947	2.0147

$$\underbrace{\qquad\qquad}_{236.0526} \qquad\qquad \underbrace{\qquad\qquad}_{235.8454}$$

The masses of the particles involved are given in atomic mass units (amu) with 1 amu being equal to $1.660\ 540\ 2 \times 10^{-27}$ kg. The equation shows a loss of (236.0526 – 235.8454), i.e. 0.2072 amu or $3.440\ 639 \times 10^{-26}$ kg. That means the liberation of $3.092\ 446\ 3 \times 10^{-11}$ J for the fission of 1 atom of $^{235}_{92}$U. The fission of 1 gram ($25.626\ 1 \times 10^{20}$ atoms) would, therefore, release $7.924\ 733\ 8 \times 10^{10}$ J.

Notes

Where a short title is provided, more details are given in the bibliography on pp. 247–8.

1. The Invention of Gunpowder

1. Thucydides, *History of the Peloponnesian War*, trans. Rex Warner (Penguin Books, 1954), p. 172
2. Ibid., p. 325
3. A six-page *Book of Fires* (Liber Ignum), attributed to Marcus Graecus (Mark the Greek), lists thirty-five recipes. Partington (Ch. II) and Hime, *Artillery* (Ch. III) give more detail
4. Ibid., recipe no. 3
5. Ibid., recipe no. 26
6. Partington, p. 19
7. Needham's *Priestley Lecture*, p. 314.
8. *School Science Review*, vol. XLVII, no. 163, June 1966, pp. 630–1
9. Partington, p. 66
10. Davis, Article in *Industrial and Engineering Chemistry*, vol. 20, no. 7, July 1928, p. 773
11. Ibid., p. 774
12. Ibid., p. 776
13. Partington, p. 78
14. Ibid., Preface, p. xii
15. Edward Gibbon, *The Decline and Fall of the Roman Empire* (Methuen & Co., 1912), vol. vi, p. 10
16. Ibid., vol. vii, p. 85

2. Making Gunpowder

1. John Bate, *The Mysteries of Nature and Art*, published in London in the seventeenth century
2. T.M. Lowry, *Historical Introduction to Chemistry* (Macmillan & Co., 1915), p. 5
3. Partington, p. 309
4. Read, p. 72
5. Gösta E. Sandström, *The History of Tunnelling* (Barrie & Rockliff, 1963), p. 277

6. T.E. Thorpe, *Essays in Historical Chemistry* (Macmillan & Co., 1894), pp. 89–90
7. M.B. Donald, 'History of the Chile Nitrate Industry – I', *Annals of Science*, vol. 1, 1936, p. 29
8. Quoted by M.E. Weeks, 'The Discovery of the Elements', *Journal of Chemical Education*, 1945, p. 12
9. From a speech by Frasch when accepting the Perkin Medal in 1912. Reported in *Industrial Engineering Chemistry*, 4, 131 (1912)

3. The Powder Trust

1. Israel Putnam, *The Oxford Dictionary of Quotations* (Oxford, 1964), p. 404
2. Wilkinson, *Explosives in History*, p. 16
3. Wilkinson, *Lammot du Pont*, p. 61
4. Ibid., p. 283
5. Ibid., p. 45
6. Dutton, p. 173

4. Testing Gunpowder

1. R. Coleman, *The Philosophical Magazine*, vol. ix, 1801, p. 360
2. *The Complete Works of Count Rumford*, American Academy of Arts and Sciences, vol. II, p. 98
3. *Collected Works of Count Rumford*, ed. Sanborn C. Brown (Harvard University Press, 1970), p. 451
4. *Dictionary of National Biography*, vol. xix, p. 688
5. Needham, *Priestley Lecture*, p. 334
6. Ibid., p. 335
7. Ibid., p. 335
8. J.D. Bernal, *Science in History* (London, C.A. Watts, 1954), pp. 238–9

5. 'Crakys of War'

1. Hugh Ross Williamson, *The Gunpowder Plot* (Faber & Faber, 1951), p. 109
2. Ibid., p. 66
3. Ibid., p. 206
4. Partington, p. 239
5. Hime, *Origin of Artillery*, p. 113
6. Partington, p. 49
7. Needham, *Priestley Lecture*, p. 320
8. General Cavalié Mercer, *Journal of the Waterloo Campaign* (London, Peter Davies Ltd, 1927), p. 153
9. Ibid., p. 92
10. Hime, *Origin of Artillery*, p. 128
11. A.W. Wilson, *The Story of the Gun* (Royal Artillery Institution, 1985)
12. Ibid., p. 8
13. Winston S. Churchill, *A History of the English Speaking Peoples*, vol. 1 (Cassell & Co.), p. 279
14. Colin Martin and Geoffrey Parker, *The Spanish Armada* (Hamish Hamilton, 1988), p. 13
15. Garrett Mattingly, *The Defeat of the Spanish Armada* (Jonathan Cape, 1959), p. 271
16. Martin and Parker, p. 35
17. Mattingly, p. 304
18. Ibid., p. 285
19. Ibid., p. 106
20. *Henry V*, part III, Prologue
21. Churchill, p. 272
22. Valentine Blacker (1778–1823), 'Oliver's Advice', *Oxford Dictionary of Quotations*
23. H.C.B. Rogers, *Weapons of the British Soldier* (London, Seeley Service & Co. Ltd), Imperial services library, vol. 5, p. 108

6. Mining and Civil Engineering

1. Robert Galloway, *Annals of Coal Mining* (Newton Abbot, David & Charles, 1971), p. 349
2. Earl, p. 16
3. John Ayrton Paris, *The Life of Sir Humphry Davy* (1831), p. 300
4. Robert Galloway, *A History of Coal Mining in Great Britain* (Newton Abbot, David & Charles, 1969), p. 160
5. Ayrton Paris, p. 78
6. Galloway, *History*, p. 164
7. John Davy, *Collected Works of Sir Humphry Davy*, vol. 1 (London, Smith, Elder & Co., 1839), p. 11
8. Galloway, *History*, p. 167
9. John Davy, *Collected Works of Sir Humphry Davy*, vol. 1, p. 400
10. *The Newcastle Magazine* (1828), vol. vii, p. 9
11. L.T.C. Rolt, *Isambard Kingdom Brunel* (Longmans, Green & Co., 1957), p. 136
12. Henry de Mosenthal, Article in *Journal of the Society of Chemical Industry* (May 1899), p. 446

7. Gunpowder Modifications

1. Hime, *Origin of Artillery*, p. 168
2. Bergengren, p. 171

8. Nitroglycerine

1. Macdonald, pp. 161–2
2. Ettore Molinari, *Treatise on General and Industrial Organic Chemistry* (J. & A. Churchill, 1913), p. 223
3. Philipp Lenard, *Great Men of Science* (G. Bell & Sons Ltd, 1933), p. 140
4. E.J. Holmyard, *Makers of Chemistry* (Oxford University Press, 1931), p. 189
5. T.M. Lowry, *Historical Introduction to Chemistry* (Macmillan & Co., 1915), p. 13
6. E.H. Tripp and S.W. Cheveley, *A Century of Fertiliser Progress* (The Dangerfield Printing Co. Ltd, 1939), p. 30
7. 'The Story of Fritz Haber', *Science and Technology in Society*, 2. SATIS no. 207, Association for Science Education, p. 3
8. Eduard Farber (ed.), *Great Chemists* (New York and London, Interscience Publishers, 1961), p. 1310
9. Ibid., p. 1311
10. L.F. Haber, *The Poisonous Cloud* (Oxford, Clarendon Press, 1986), p. 1
11. Bergengren, p. 9
12. British patent, no. 1345, 1867
13. Bergengren, p. 24
14. Ibid., p. 28

9. Dynamite

1. Bergengren, p. 50
2. Ibid., p. 175
3. Ibid., p. 175
4. F.D. Miles, *A History of Research in the Nobel Division of ICI* (ICI Nobel Division, 1955). Quoted in Bergengren, p. 92
5. Dutton, p. 141

6. W.H. Carr, *The du Ponts of Delaware* (Frederick Muller, 1965), p. 201
7. Bergengren, p., 121
8. D.B. Barton, *History of Tin Mining and Smelting in Cornwall* (D. Bradford Barton Ltd, 1965), p. 181
9. Ibid., p. 181
10. Extract from *The Secret of Machines* by Rudyard Kipling
11. Bergengren, p. 124
12. Ibid., p. 155
13. Ibid., p. 170
14. Ibid., p. 132
15. Ibid., p. 175
16. Ibid., p. 177
17. Ibid.
18. Ibid., p. 157
19. Ibid., p. 202
20. Ibid., p. 190
21. Ibid., p. 195
22. Ibid., p. 137
23. Reader, p. 17

10. Guncotton

1. Ralph E. Oesper, 'C.F. Schünbein; Part 1. Life and Character', *Journal of Chemical Education*, March, 1929, p. 439
2. Ibid., p. 432
3. Ibid., p. 432
4. Macdonald, p. 15
5. Oesper, p. 434
6. Ibid., p. 432
7. British patent, no. 11407, 1846
8. Quoted in *Industrial Archaeology*, 1968, p. 17
9. Macdonald, pp. 48–9
10. Ibid., pp. 49–50
11. Ibid., p. 51
12. Oesper, *Journal of Chemical Education*, April 1929, p. 685
13. British patent, no. 320, 1862
14. British patent, no. 1102, 1865

11. Smokeless Powders

1. British patent, no. 1471, 1888
2. Bergengren, p. 106
3. Ibid., p. 113
4. British patents, no. 5614, 2 April 1889 and no. 11664, 22 July 1889
5. Reader, p. 142
6. Bergengren, p. 116
7. Reader, p. 143
8. Ibid., p. 145
9. Ibid., p. 381

12. Lyddite and TNT

1. David Lloyd George, *War Memoirs* (Ivor Nicholson and Watson, 1933), p. 198
2. Ibid., p. 211
3. Ibid., p. 125
4. Ibid., p. 554
5. Ibid., p. 209
6. Ibid., p. 617
7. Ibid., p. 589
8. Ibid., p. 589
9. Dick Dent, 'Famous Men Remembered', *The Chemical Engineer*, December 1986, p. 56
10. Lloyd George, pp. 577–80
11. Ibid., p. 128
12. Ibid., p. 597
13. *Hamlet*, III, iv
14. Norman Gladden, *Ypres, 1917, A Personal Account* (William Kemble & Co. Ltd, 1967), p. 61
15. Winston S. Churchill, *World Crisis*, part II, 1916–18 (Thornton Butterworth Ltd, 1927), p. 318

13. Setting It Off

1. Basil Valentine, *Collected Writings*, 3rd German edn (Hamburg, 1700), p. 289
2. Pepys's diary, 11 November 1663
3. Ibid., 14 March 1662
4. *Philosophical Transactions of the Royal Society*, 1800, vol. XC, p. 204
5. Macdonald, *Historical Papers*, p. 2
6. British patent, no. 3032, 1807
7. Maj-Gen. Sir Alexander John Forsyth, *The Reverend Alexander John Forsyth* (Aberdeen University Press, 1909), p. 21
8. *The Rise and Progress of the British Explosives Industry* (Whittaker and Co.), p. 117

14. Nuclear Fission

1. Sir W.H. Ramsay, *The Death Knell of the Atom*, 1905. Reported in *The News Edition of Industrial and Chemical Engineering*, vol. 8, no. 2, 20 January 1930, p. 18
2. George B. Kauffman, ed., *Frederick Soddy* (D. Reider Publishing Co.), p. 182
3. A.S. Eve, *Rutherford* (Cambridge University Press, 1939), p. 102
4. Eduard Farber, *Nobel Prize Winners in Chemistry* (Abelard-Schuman, 1953), p. 182

5. R. Jungk, *Heller als tausend Sonnen* (Penguin Books), p. 71
6. S.R. Weart and G.W. Szilard (eds), *Leo Szilard: His Version of the Facts* (MIT Press, 1978), p. 55
7. Clark, p. 214
8. Winston S. Churchill, *The Second World War*, vol. III, 'The Grand Alliance' (Cassell & Co. Ltd), p. 730
9. Jane Wilson (ed.), 'All in Our Time', *Bulletin of the Atomic Scientists*, 1975, p. 95
10. Eugene P. Wigner, *Symmetries and Reflections* (Indiana University Press, 1967). Reprinted by Oxbow Press, 1979, p. 240
11. A.H. Compton, *Atomic Quest* (Oxford University Press, 1956), p. 144
12. Weart and Szilard, p. 146

15. The Manhattan Project

1. P. Goodchild, *J. Robert Oppenheimer: Shatterer of Worlds* (Houghton Mifflin, 1980), p. 56
2. Libby, p. 102
3. Groves, p. 63
4. Ibid., p. 96
5. J.W. Kunetka, *Oppenheimer; The Years of Risk* (Prentice-Hall, 1982), p. 47
6. Goodchild, p. 79
7. McKay, p. 57
8. Kunetka, p. 68
9. H.G. Graetzer and D.L. Anderson, *The Discovery of Nuclear Fission* (Van Nostrand Reinhold, 1971), p. 106
10. Jane Wilson (ed.), 'All in Our Time', p. 230
11. Groves, p. 298
12. Goodchild, p. 162

13. Ibid., p. 162
14. Graetzer and Anderson, p. 108–9
15. Winston S. Churchill, *The Second World War*, vol. VI, 'Triumph and Tragedy' (Cassell & Co. Ltd, 1953), p. 553
16. Marshall, p. 232

16. Nuclear Fusion

1. Rhodes, *The Making of the Atom Bomb*, p. 777
2. Rhodes, *Dark Sun*, p. 205
3. Jungk, p. 257
4. Ibid., pp. 265–6
5. Ibid., p. 286
6. US Atomic Energy Commission, *In the Matter of J. Robert Oppenheimer: Transcript of Hearing Before Personal Security Board* (MIT Press, 1954), p. 712
7. Ibid., p. 710
8. Ibid., p. 710
9. Ibid., p. 726
10. Ibid., p. 137
11. Rhodes, *Dark Sun*, pp. 556–7
12. USA EC Statement issued on 29 June 1954, pp. 51–2 (1049–50)
13. G. Kennan, *Contribution to Memorial Service for J. Robert Oppenheimer*, 25 February 1967. Box 43, Oppenheimer Papers, US Library of Congress, Washington, DC
14. Rhodes, *Dark Sun*, p. 577
15. David Hawkins, *Manhattan District History, Project Y* (The Los Alamos Project, Los Alamos Scientific Library, 1947), p. 294
16. Rhodes, *Looking for America* (1979), p. 116

Bibliography

Akhavan, J. *The Chemistry of Explosives* (The Royal Society of Chemistry, 1998)

Antonioli, G. and Masera, G. *Explosives* (Milan, Italesplosivi SpA, 1982)

Bailey, A. and Murray, S.G. *Explosives, Propellants, and Pyrotechnics* (Brassey's (UK), 1989)

Barton, D.B. *History of Tin Mining and Smelting in Cornwall* (D. Bradford Barton Ltd, 1965; Cornwall Books, 1989)

Bergengren, E. *Alfred Nobel, The man and his work* (English translation by A. Blair, Thomas Nelson & Sons, Edinburgh, 1962)

Bibby, Leona M. *The Uranium People* (New York, Crane Russack & Charles Scribner's Sons, 1979)

Brock, A.St.H. *A History of Fireworks* (Harrap, 1949)

Buchanan, Brenda J. (ed.) *Gunpowder; The History of an International Technology* (Bath University Press, 1996)

Cartwright, A.P. *The Dynamite Company* (Macdonald, 1964)

Chidsey, D.B. *Goodbye to Gunpowder* (New York, Crown Publishers, Inc., 1963)

Clark, R.W. *The Greatest Power on Earth* (Sidgwick & Jackson, 1980)

Cocroft, Wayne. *Oare Gunpowder Works* (The Faversham Society, 1994)

Cook, M.A. *The Science of High Explosives* (Chapman & Hall, 1958)

——. *The Science of Industrial Explosive* (Utah USA, IRECO Chemicals, Salt Lake City, 1974)

Crocker, Glenys. *The Gunpowder Industry* (Shire Publications Ltd, 1986)

——. *Gunpowder Mills Gazetteer* (The Wind and Watermill Section of The Society for the Protection of Ancient Buildings, 1988)

Davis, T.L. *The Chemistry of Powder and Explosives* (Wiley, 1943)

Dolan, John E. and Oglethorpe, Miles K. *Explosives in the Service of Man* (The Royal Commission on the Ancient and Historical Monuments of Scotland, 1996)

Dutton, W.S. *One Thousand Years of Explosives* (New York, Holt, Rinehart & Winston, Inc., 1960)

Earl, B. *Cornish Explosives* (The Trevithick Society, Cornwall, 1978)

Fermi, L. *Atoms in the Family: my life with Enrico Fermi* (Chicago University Press, 1954)

Fordham, S. *High Explosives and Propellants* (Pergamon Press, 1966)

Gibbs, T.R. and Popolato, A. *LASL Explosive Properties Data* (University of California Press, 1980)

Goad, K.J.W. and Archer, E. *Ammunition* (Pergamon Press, Oxford, 1990)

Goodchild, P.J. *Robert Oppenheimer* (London, BBC Publications, 1989)

Gowing, M. *The Development of Atomic Energy: Chronology of Events, 1939–1978* (UK Atomic Energy Authority, London, 1979)

Groueff, S. *Manhattan Project* (Boston, Little, Brown & Co., 1967)

Groves, L.R. *Now it Can be Told* (Harper & Brothers, 1962)

Guttmann, O. *Blasting* (C. Griffin, 1906)

——. *Manufacture of Explosives, Twenty Years Progress* (Whittaker, 1909)

Hime, H.W.L. *Gunpowder and Ammunition: Their Origin and Progress* (London, 1904)

——. *The Origin of Artillery* (Longman, Green & Co., 1915)

Jungk, R. *Heller als tausend Sonnen* (Alfred Scherz Verlag, 1956) [*Brighter than a Thousand Suns*, English translation by J. Cleugh (Gollancz & Hart-Davis, 1958; Penguin, 1960)]

Kelleher, G.D. *Gunpowder to Guided Missiles, Ireland's War Industries* (John F. Kelleher, Inniscarra, Co. Cork, Ireland, 1993)

Köhler, J. and Meyer, R. *Explosives*, 7th edn (Weinheim, VCH-Verlagsges, Weinheim, 1993)

Libby, Leona Marshall. *The Uranium People* (Crane Russack, 1979)

McAdam, R. and Westwater, R. *Mining Explosives* (Oliver & Boyd, Edinburgh, 1958)

MacDonald, George W. *Historical Papers on Modern Explosives* (London and New York, Whittaker & Co., 1912)

McKay, A. *The Making of the Atomic Age* (Oxford University Press, 1984)

Major, J. *The Oppenheimer Hearing* (New York, Stein & Day, 1971)

Marshall, A. *Explosives. Their Manufacture, Properties, Tests and History* (J. & A. Churchill, 1915)

Miles, F.D. *A History of Research in the Nobel Division of ICI* (ICI, 1955)

Needham, Joseph. *Science and Civilisation in China*, vol. V (Cambridge University Press, 1976)

——. *Gunpowder as the Fourth Power, East and West* (Hong Kong University Press, 1985)

——. *The Priestly Lecture, 1983. The Proceedings of the Third BOC Priestly Conference* (London, The Royal Society of Chemistry, 1983)

Partington, J.R. *A History of Greek Fire and Gunpowder* (Cambridge, Heffer, 1960; New York, Barnes & Noble, Inc., 1960)

Patterson, Edward M. *Gunpowder, Terminology and Incorporation* (The Faversham Society, 1986)

——. *Gunpowder Manufacture at Faversham; Oare and Marsh Factories* (The Faversham Society, 1986)

Percival, Arthur J. *The Great Explosion at Faversham, 1916* (reprinted in *Archaeologia Cantiana*, vol. C (1985))

——. *Faversham's Gunpowder Industry and its Development* (The Faversham Society, 1967)

Plimpton, G. *Fireworks, A History and Celebration* (Doubleday, 1984)

Read, J. *Explosives* (Pelican Books, 1942)

Reader, W.J. *Imperial Chemical Industries – A History* (Oxford University Press, 1970)

Rhodes, R. *The Making of the Atomic Bomb* (Simon & Schuster, 1986)

——. *Dark Sun: the Making of the Hydrogen Bomb* (Simon & Schuster, 1995)

Schück, H. and Sohlman, R. *The Life of Alfred Nobel* (William Heinemann, 1929)

Simmons, W.H. *A Short History of the Royal Gunpowder Factory at Waltham Abbey* (Explosives Research and Development Establishment, 1963)

Smyth, H.D. *Atomic Energy* (US Government Printing Office, 1945)

Sohlman, R. *The Legacy of Alfred Nobel – The Story Behind the Nobel Prizes* (Bodley Head, 1983)

Szilard, L. *Selected Recollections and Correspondence*, edited by S.R. Weart and G.W. Szilard (Cambridge, MIT Press, 1972)

Taylor, W. *Modern Explosives* (London, The Royal Institute, 1959)

The Rise and Progress of the British Explosives Industry, VIIth International Congress of Applied Chemistry (Whittaker, 1909)

The Royal Gunpowder Factory, Waltham Abbey, Essex – An RCHME Survey, 1993 (RCHME, 1994)

Urbanski, T. *Chemistry and Technology of Explosives*, 4 vols, translated by I. Jeczalikowa and S. Laverton (Oxford, Pergamon Press, 1964–85)

Van Gelder, A.P. and Schlatter, H. *History of the Explosives Industry in America* (Columbia University Press, 1927)

Wilkinson, N.B. *Lammot du Pont and the American Explosives Industry, 1850–1884* (University Press of Virginia, 1984)

——. *Explosives in History* (The Hagley Museum, Wilmington, Delaware, 1966)

York, H. *The Advisors: Oppenheimer, Teller and the Super Bomb* (San Francisco, 1976)

Index